物联网开发与应用丛书

面向物联网的Android应用开发与实践

廖建尚 张凯 郝丽萍 编著

電子工業出版社
Publishing House of Electronics Industry
北京•BEIJING

内 容 简 介

本书主要介绍物联网系统中的 Android 应用开发。全书先进行理论知识的学习，深入浅出地介绍 Java 开发基础、Android 开发基础和 Android 开发进阶等理论知识，然后进行实际案例的开发，最后进行开发验证和总结拓展，将理论学习和开发实践紧密结合起来。每个实际案例均给出完整的开发代码和配套 PPT，读者可以在此基础上快速地进行二次开发。

本书既可作为高等学校相关专业的教材或教学参考书，也可供相关领域的工程技术人员参考。对于 Android 应用开发和物联网系统开发的爱好者来说，本书也是一本贴近实际应用的技术读物。

本书配有教学课件和开发代码，读者可登录华信教育资源网（www.hxedu.com.cn）免费注册后下载。

图书在版编目（CIP）数据

面向物联网的 Android 应用开发与实践 / 廖建尚，张凯，郝丽萍编著. —北京：电子工业出版社，2020.8
（物联网开发与应用丛书）
ISBN 978-7-121-39422-5

Ⅰ. ①面… Ⅱ. ①廖… ②张… ③郝… Ⅲ. ①移动终端－应用程序－程序设计 Ⅳ. ①TN929.53

中国版本图书馆 CIP 数据核字（2020）第 154958 号

责任编辑：田宏峰
印　　刷：北京天宇星印刷厂
装　　订：北京天宇星印刷厂
出版发行：电子工业出版社
　　　　　北京市海淀区万寿路 173 信箱　邮编：100036
开　　本：787×1 092　1/16　印张：19.5　字数：500 千字
版　　次：2020 年 8 月第 1 版
印　　次：2024 年 7 月第 10 次印刷
定　　价：79.00 元

凡所购买电子工业出版社图书有缺损问题，请向购买书店调换。若书店售缺，请与本社发行部联系，联系及邮购电话：（010）88254888，88258888。

质量投诉请发邮件至 zlts@phei.com.cn，盗版侵权举报请发邮件至 dbqq@phei.com.cn。

本书咨询联系方式：tianhf@phei.com.cn。

FOREWORD 前言

近年来，物联网、移动互联网、大数据和云计算的迅猛发展，逐步改变了社会的生产方式，大大提高了生产效率和社会生产力。工业和信息化部发布的《信息通信行业发展规划物联网分册（2016—2020 年）》总结了“十二五”规划中物联网发展所获得的成就，并分析了“十三五”期间面临的形势，明确了物联网的发展思路和目标，提出了物联网发展的 6 大任务，分别是强化产业生态布局、完善技术创新体系、推动物联网规模应用、构建完善标准体系、完善公共服务体系、提升安全保障能力；提出了 4 大关键技术，分别是传感器技术，体系架构共性技术，操作系统，以及物联网与移动互联网、大数据融合关键技术；提出了 6 大重点领域应用示范工程，分别是智能制造，智慧农业，智能家居，智能交通和车联网，智慧医疗和健康养老，以及智慧节能环保；指出要健全多层次、多类型的物联网人才培养和服务体系，支持高校、科研院所加强跨学科交叉整合，加强物联网学科建设，培养物联网复合型专业人才。该发展规划为物联网发展指出了一条鲜明的道路，同时也表明了我国在推动物联网应用方面的坚定决心，相信物联网的规模会越来越大。

本书采用任务式开发的学习方法，利用实际的开发案例，由浅入深地介绍 Android 应用开发技术，每个案例均有完整的开发过程，即生动的开发场景、明确的开发目标、深入浅出的原理学习、详细的开发实践、任务验证、开发小结、思考与拓展。每个案例均附有完整的开发代码，读者在此基础上可以快速地进行二次开发，从而将其转化为各种比赛和创新创业的案例。

本书分为 4 篇。

第 1 篇为 Java 开发基础。本篇主要介绍 Java 的基本知识和语法，主要内容包括 Android 应用开发环境、Java 开发基础、海拔高度数据的定义与转换、温度和湿度数据的记录、智能背包系统的设计。

第 2 篇为 Android 开发基础。本篇主要介绍 Android 的基本知识和语法，主要内容包括 Android 项目框架与调试、空气质量显示界面的设计、城市气象监控设备管理系统应用界面的设计、城市环境系统功能菜单的设计、城市灯光控制系统界面事件的处理。

第 3 篇为 Android 开发进阶。本篇主要介绍 Android 的高级知识和语法，主要内容包括工厂通风系统界面的切换、工厂火警监测系统界面的设计、设备列表管理界面的设计、智能电表日志的记录、光照度记录的查询、智能医疗仪表图形的动态显示、远程控制服务端的通信。

第 4 篇为物联网 Android 应用开发。本篇主要以实例的形式讲述面向物联网系统的 Android 应用程序开发，主要内容包括物联网系统框架及 Android 开发接口、仓库环境管理系统的设计。

本书既可作为高等学校相关专业的教材或教学参考书，也可供相关领域的工程技术人员参考。对于物联网开发和 Android 开发的爱好者来说，本书也是一本深入浅出的技术读物。

本书在编写过程中借鉴和参考了国内外专家、学者和技术人员的相关研究成果，作者尽可能按学术规范予以说明，但难免会有疏漏，在此向相关作者表示深深的敬意和谢意。如有疏漏，请及时通过出版社与作者联系。

本书的出版得到了广东省自然科学基金项目（2018A030313195）和广州市科技计划项目（201804010262）的资助。感谢中智讯（武汉）科技有限公司在本书编写过程中提供的帮助，特别感谢电子工业出版社在本书出版过程中给予的大力支持。

由于本书涉及的知识面广，限于作者的水平和经验，疏漏之处在所难免，恳请广大专家和读者批评指正。

作　者

2020 年 6 月

CONTENTS 目录

第1篇　Java 开发基础

第 2 篇 Android 开发基础

第 3 篇 Android 开发进阶

第 1 篇

Java 开发基础

本篇主要介绍 Java 的基本知识和语法，通过开发实践帮助读者熟悉 Java 的开发基础。本篇共有 5 个任务：

任务 1 为 Android 应用开发环境。

任务 2 为 Java 开发基础。

任务 3 为海拔高度数据的定义与转换。

任务 4 为温度和湿度数据的记录。

任务 5 为智能背包系统的设计。

任务 1

Android 应用开发环境

本任务通过介绍 Android 系统架构、Android 应用开发框架和 Android Studio 的安装部署，帮助读者熟练掌握 Android 应用程序创建及运行的方法。

1.1 开发场景：如何搭建 Android 应用开发环境

Android 应用开发环境（Android Studio IDE）是基于 IntelliJ IDEA 构建的，如图 1.1 所示。通过本任务的学习，读者可以搭建 Android 应用开发环境，创建第一个 Android 应用程序。

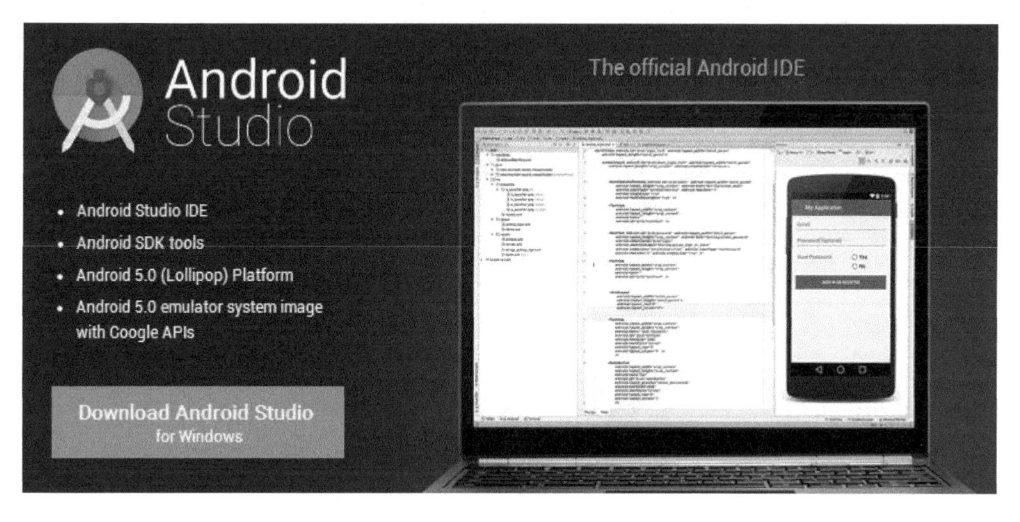

图 1.1 Android 应用开发环境

1.2 开发目标

（1）知识目标：了解 Android 系统架构，熟悉 Android 应用开发框架。

（2）技能目标：熟悉 Android Studio 的安装与配置，熟悉 Android 应用程序的创建与运行，熟悉 Android 项目的导入。

（3）任务目标：通过学习 Android 系统架构、Android 开发框架和 Android Studio 的安装部署，熟练掌握 Android 应用程序的创建、运行及项目导入方法。

1.3　原理学习：Android 系统架构和开发框架

1.3.1　Android 系统架构

Android 系统架构如图 1.2 所示

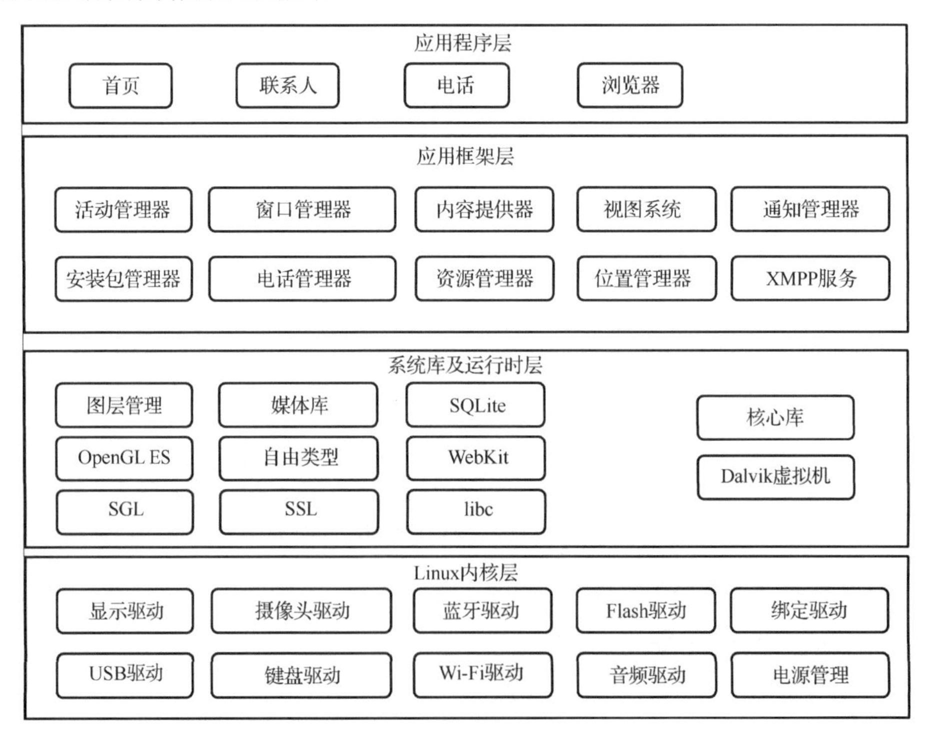

图 1.2　Android 系统架构

Android 系统架构和其操作系统架构一样，都采用了分层架构。Android 系统架构共分四层，分别是应用程序层、应用框架层、系统库及运行时层和 Linux 内核层。

（1）应用程序层：该层提供核心应用程序包，如首页、联系人、电话和浏览器等，开发者可以设计和编写相应的应用程序。

（2）应用框架层：该层是 Android 应用开发的基础，包括活动管理器、窗口管理器、内容提供器、视图系统、通知管理器、安装包管理器、电话管理器、资源管理器、位置管理器和 XMPP 服务。

（3）系统库及运行时层。系统库中的库文件主要包括图层管理、媒体库、SQLite、OpenGL ES、自由类型、WebKit、SGL、SSL 和 libc；运行时包括核心库和 Dalvik 虚拟机。核心库不仅兼容大多数 Java 所需要的功能函数，还包括 Android 的核心库，如 android.os、android.net、 android.media 等；Dalvik 虚拟机是一种基于寄存器的 Java 虚拟机，主要完成对生命周期、堆栈、线程、安全和异常的管理，以及垃圾回收等功能。

（4）Linux 内核层。Linux 内核层提供各种硬件驱动，如显示驱动、摄像头驱动、蓝牙驱动、键盘驱动、Wi-Fi 驱动、音频驱动、Flash 驱动、绑定驱动、USB 驱动、电源管理等。

1.3.2 Android 应用开发框架

Android 应用开发包含基本的应用功能开发、数据存储和网络访问三大模块。

1. 基本的应用功能开发

1）Android 应用程序的组成

Android 应用程序一般由 Activity、Broadcast Receiver、Service、ContentProvider 组成。

（1）Activity。Activity 是最基本的模块。Android 应用程序往往包含多个 Activity 实例。一个 Activity 实例就是手机上的一屏，相当于一个网页，每个 Activity 实例运行结束后都有一个返回值。Android 应用程序会记录从首页到其他界面的跳转记录，并将以前的 Activity 实例压入系统堆栈。用户可以通过编程的方式删除堆栈中的 Activity 实例。

Activity 主要用于关联界面资源文件（“res/layout”目录下的 XML 资源，也可以不包含界面资源），其内部包含控件的显示设计、界面交互设计、事件的响应设计，以及数据处理设计、导航设计等。

（2）BroadcastReceiver。BroadcastReceiver 提供了在 Android 应用程序进程间进行通信的机制，例如，当来电时，可以通过 BroadcastReceiver 广播消息。对用户而言，BroadcastReceiver 是不透明的，用户无法看到事件，BroadcastReceiver 通过 NotificationManager 来通知用户事件发生了。BroadcastReceiver 既可以在资源 AndroidManifest.xml 中注册，也可以在代码中通过 Context.registerReceiver()函数进行注册，只要注册了，当事件来临时，即使应用程序没有启动，系统也在需要的时候自动启动应用程序。另外，应用程序可以很方便地通过 Context.sendBroadcast()将自己的事件广播给其他应用程序。

（3）Service。Service 主要用于在后台处理一些耗时的逻辑，或者执行某些需要长期运行的任务，甚至可以在应用程序退出的情况下，让 Service 在后台继续保持运行状态。Service 拥有生命周期方法，可以监控服务状态的变化，以便在合适的阶段执行工作。

（4）ContentProvider。ContentProvider 提供了在应用程序之间进行数据交换的机制。应用程序可以通过实现一个 ContentProvider 的抽象接口将自己的数据暴露出去，并隐蔽具体的数据存储实现。ContentProvider 提供了基本的 CRUD（Create、Read、Update、Delete）接口，并实现了权限机制，从而可以保护数据交互的安全性。

2）Android 应用程序的项目文件

一个标准的 Android 应用程序的项目文件包含以下几个部分：

（1）src 目录：该目录用于保存 Java 代码部分（包含 Activity）。

（2）R.java 文件：这个文件是由 Eclipse 自动生成与维护的，用户不需要修改，该文件提供了 Android 应用程序资源的全局索引。

（3）Android Libraries 目录：该目录用于保存运行应用程序的 Android 库。

（4）assets 目录：该目录用于保存多媒体等文件。

（5）res 目录：该目录用于保存资源文件，和 VC 中的资源目录类似。该目录中的 drawable 保存的是图片文件，layout 保存的是布局文件，values 保存的是 strings.xml（字符串）、colors.xml（颜色）和 arrays.xml（数组）资源。

（6）AndroidManifest.xml：这个文件非常重要，是整个应用程序的配置文件。在这个文件中，需要声明所有要用到的 Activity、Service、Receiver 等。

2. 数据存储

Android 的数据存储方式包括 SharedPreferences、文件存储、SQLite 数据库存储、内容提供器（ContentProvider）以及网络存储，具体如下：

（1）SharedPreferences。SharedPreferences 是由 Android 提供的一种最简单的数据存储方式，默认存在应用程序项目文件的“data/<package name>/shared_prefs”中，通过 getSharedPreferences 函数可获取 SharedPreferences 对象并进行读写操作。

（2）文件存储。通过 openFileInput、openFileOutput 等系统提供的 API 可以进行数据的读写访问。特别需要注意的是，在 Android 中，应用程序的数据是私有的，也就是说，其他应用程序无法访问当前应用程序所产生的文件。

（3）SQLite 数据库存储。SQLite 数据库存储是通过继承 SQLiteOpenHelper 类并获取此类的应用程序级别的实例来进行数据库操作的。SQLiteOpenHelper 类中提供了默认的 CRUD 接口，便于对应用程序进行数据存储操作。

（4）ContentProvider。ContentProvider 是通过调用其他应用程序的数据接口来实现数据存储的。

（5）网络存储。网络存储是通过网络访问服务接口（如 WebService 数据访问接口）来实现数据存储的。

3. 网络访问

网络访问主要是对 HTTP 访问技术的封装，并通过 java.net.*以及 Android.net.*提供的 HttpPost、DefaultHttpClient、HttpResponse 等类提供的访问接口来实现 Web 服务访问。

1.4　开发实践：创建第一个 Android 应用程序

Android 应用程序的开发工具都是开源的，可以从网上下载。在编写 Android 应用程序之前需要的开发工具包括 JDK、Android SDK 和 Android Studio。

1. JDK

JDK（Java Development Kit）包含 javac（Java 编译器）、JRE（Java 运行时环境）、Java 类库等。其中 JRE 包含 JVM（Java 虚拟机）和运行 Java 程序的环境变量支持。JDK 是 Java 的软件开发工具包，主要用于开发移动设备、嵌入式设备上的 Java 应用程序。JDK 是整个 Java 开发的核心。JDK 的构成如图 1.3 所示。

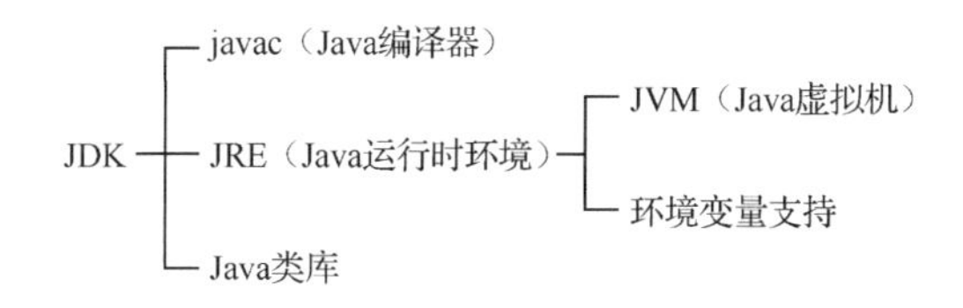

图 1.3　JDK 的构成

2. Android SDK

Android 软件开发工具包（Software Development Kit，SDK）提供了开发 Android 应用程序所需的 API 库，是构建、测试和调试 Android 应用程序所需的开发工具。

由于 Android SDK 采用 Java 语言，所以需要先安装 JDK 5.0 及以上版本。Android SDK 不用安装，将下载后的 Android SDK 压缩包解压到适当的位置即可。

3. Android Studio

Android Studio 是基于 IntelliJ IDEA 搭建的 Android 开发环境，类似于 Eclipse ADT。Android Studio 提供了集成的 Android 应用程序开发工具，用于开发和调试 Android 应用程序。

1.4.1　JDK 的安装与配置

1. 下载 JDK

读者可以在 Oracle 官网（http://www.oracle.com/technetwork/java/javase/downloads/index.html）下载 JDK 的安装包。在下载时需要注意，要根据自己计算机中的操作系统选择相应的版本。JDK 安装包的下载界面如图 1.4 所示，选择 JDK 安装包的界面如图 1.5 所示。

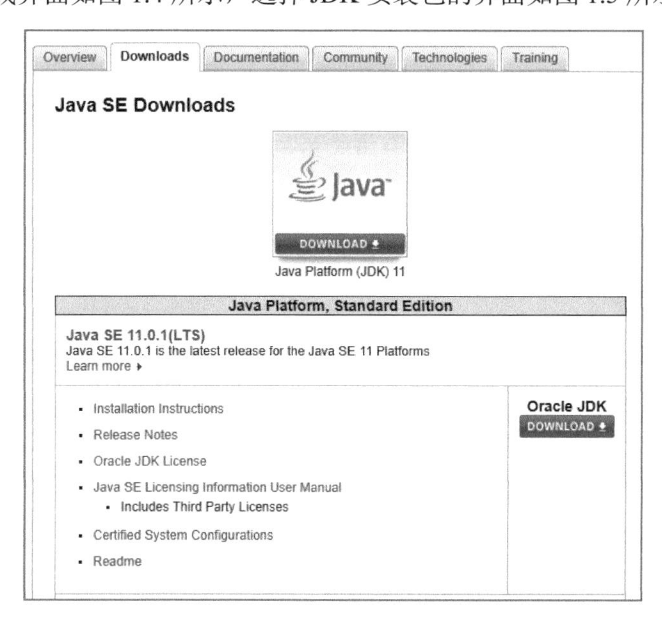

图 1.4　JDK 安装包的下载界面

Java SE Development Kit 11.0.1

You must accept the Oracle Technology Network License Agreement for Oracle Java SE to download this software.

○ Accept License Agreement　◉ Decline License Agreement

Product / File Description	File Size	Download
Linux	147.4 MB	jdk-11.0.1_linux-x64_bin.deb
Linux	154.09 MB	jdk-11.0.1_linux-x64_bin.rpm
Linux	171.43 MB	jdk-11.0.1_linux-x64_bin.tar.gz
macOS	166.2 MB	jdk-11.0.1_osx-x64_bin.dmg
macOS	166.55 MB	jdk-11.0.1_osx-x64_bin.tar.gz
Solaris SPARC	186.8 MB	jdk-11.0.1_solaris-sparcv9_bin.tar.gz
Windows	150.98 MB	jdk-11.0.1_windows-x64_bin.exe
Windows	170.99 MB	jdk-11.0.1_windows-x64_bin.zip

图 1.5　选择 JDK 安装包的界面

2. 安装 JDK

安装 JDK 时需要选择安装目录，在安装过程中会出现两次选择安装目录的提示。第一次出现在安装 JDK 时，第二次出现在安装 JRE 时。如果无特殊要求，则可按照默认的设置安装，一直到安装完毕为止。JDK 的安装配置如图 1.6 所示。

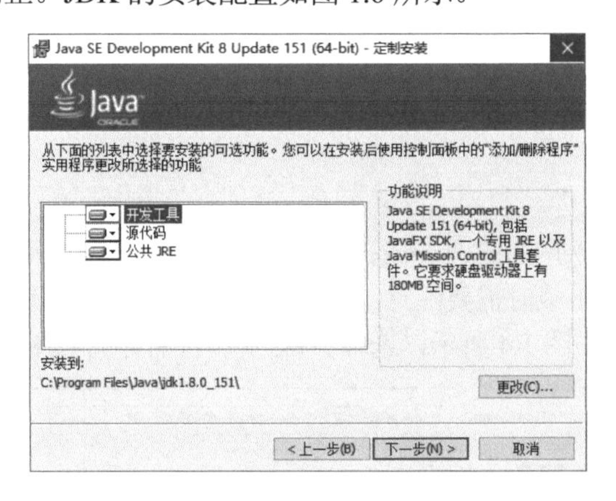

图 1.6　JDK 的安装配置

若需要更改安装目录，则建议 JDK 和 JRE 安装在 Java 文件夹中的不同子文件夹中。注意：不能将 JDK 和 JRE 安装在同一个文件夹中，否则会出错。

3. 环境变量配置

安装完 JDK 后还需要配置计算机中的环境变量，其方法是（不同版本的 Windows 操作系统略有差异）：右键单击“计算机”，在弹出的快捷菜单中选择“系统属性”，然后依次选择“高级系统设置→高级→环境变量”。Windows 10 操作系统的环境变量界面如图 1.7 所示。

在环境变量界面中创建变量 JAVA_HOME、CLASSPATH 后，还需要修改变量值，具体如下：

（1）创建变量 JAVA_HOME，变量值为 JDK 的安装目录，如“C:\Program Files\Java\jdk1.8.0_151”。

（2）创建变量 CLASSPATH，变量值为“.;C:\Program Files\Java\jdk1.8.0_151\lib\dt.jar;C:\Program Files\Java\jdk1.8.0_151\lib\tools.jar”。

（3）修改 PATH 的变量值，将“%JAVA_HOME%\bin;%JAVA_HOME%\jre\bin;”放到原变量值前面。

图 1.7　Windows 10 操作系统的环境变量界面

环境变量界面如图 1.8 所示。

图 1.8　环境变量界面

4．检验安装及配置情况

单击 Windows 操作系统的开始菜单，在“搜索程序和文件”中输入“CMD”，在弹出的命令行窗口中输入下面的命令：

```
java -version                    //该命令用于检查 Java 安装版本
javac -version                   //该命令用于检查 javac 安装版本
```

如果出现如图 1.9 所示信息，则说明环境变量配置成功（该图中的版本号是一个示例，需要检查和实际的 Java 版本号是否一致），否则需要检查环境变量的配置。

```
C:\>java -version
java version "1.8.0_151"
Java(TM) SE Runtime Environment (build 1.8.0_151-b12)
Java HotSpot(TM) 64-Bit Server VM (build 25.151-b12, mixed mode)

C:\>javac -version
javac 1.8.0_151
```

图 1.9　检验安装及配置情况

1.4.2　Android Studio 的安装与配置

1．下载 Android Studio

读者可以在 Android Studio 中文社区（https://developer.android.google.cn/studio/index.html）下载 Android Studio 安装包。Android Studio 安装包的下载界面如图 1.10 所示。

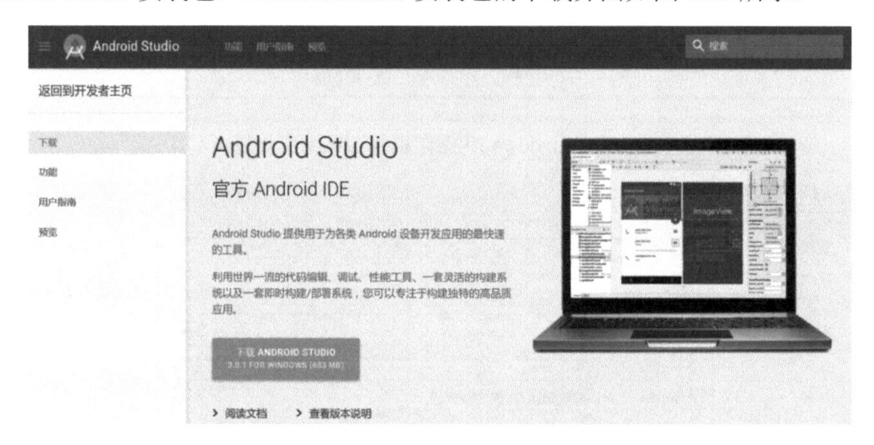

图 1.10　Android Studio 安装包的下载界面

2．安装 Android Studio

双击下载的 Android Studio 安装包可以弹出如图 1.11 所示的“Welcome to Android Studio Setup”界面，单击“Next”按钮可弹出如图 1.12 所示的“Choose Components”界面。

图 1.11　“Welcome to Android Studio Setup”界面

在“Choose Components”界面中可以选择需要安装的组件。“Android Studio”为主程序，默认已勾选；“Android Virtual Device”为 Android 虚拟设备，该组件可以在计算机中模拟安装手机的环境，可以直接在计算机中允许开发出的 Android App，也将其勾选上。继续单击“Next”按钮，弹出如图 1.13 所示的“Configuration Settings”界面。

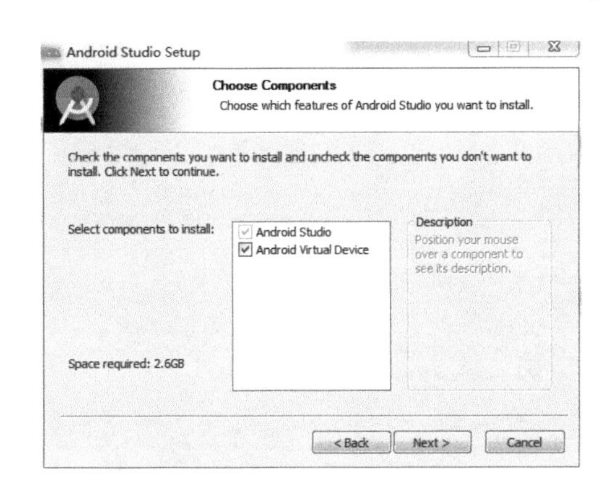

图 1.12　“Choose Components”界面

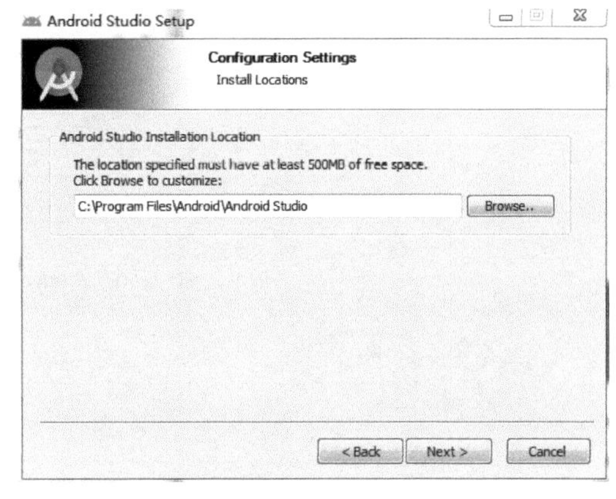

图 1.13　“Configuration Settings”界面

在“Configuration Settings”界面中单击“Browse”按钮可以选择安装的目录。继续单击“Next”按钮可以弹出如图 1.14 所示的“Choose Start Menu Folder”界面。

在“Choose Start Menu Folder”界面中可以根据需要选择是否勾选“Do not create shortcuts”选项，单击“Install”按钮可以开始 Android Studio 的安装，安装过程界面如图 1.14 所示。

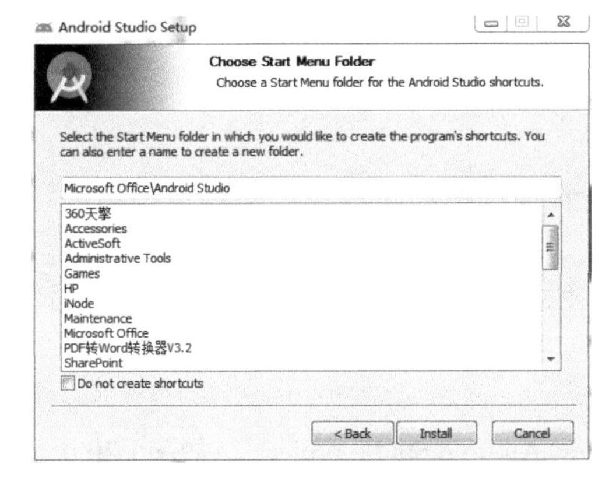

图 1.14　“Choose Start Menu Folder”界面

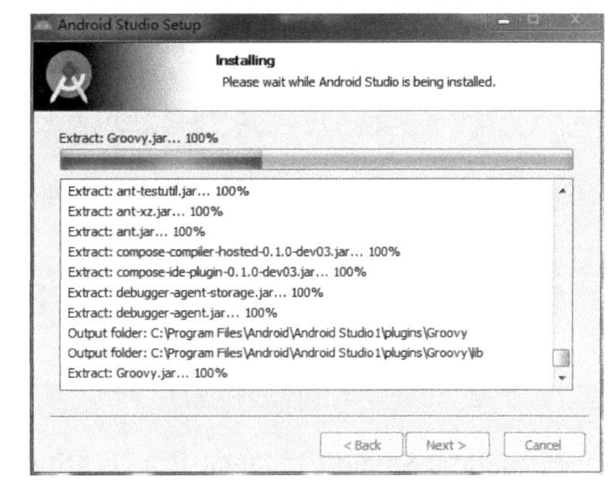

图 1.15　Android Studio 的安装过程

安装完成后勾选“Start Android Studio”，单击“Finish”按钮就可以打开 Android Studio 了。

3．配置 SDK 路径

Android Studio 启动界面如图 1.16 所示。

单击 Android Studio 启动界面右下方的“Configure”按钮，在下拉菜单中选择“Project Defaults”选项（见图 1.17），在弹出的界面中选择“Project Structure”选项（见图 1.18），此时会弹出如图 1.19 所示的“Project Structure”界面。在“Project Structure”界面中设置已经

安装好的 SDK。

图 1.16　Android Studio 启动界面

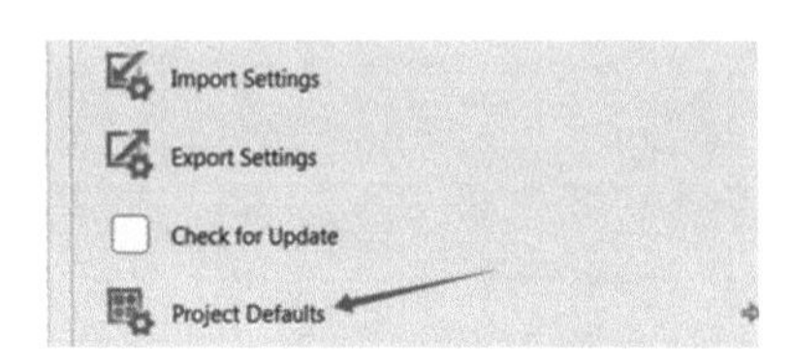

图 1.17　选择“Project Defaults”选项

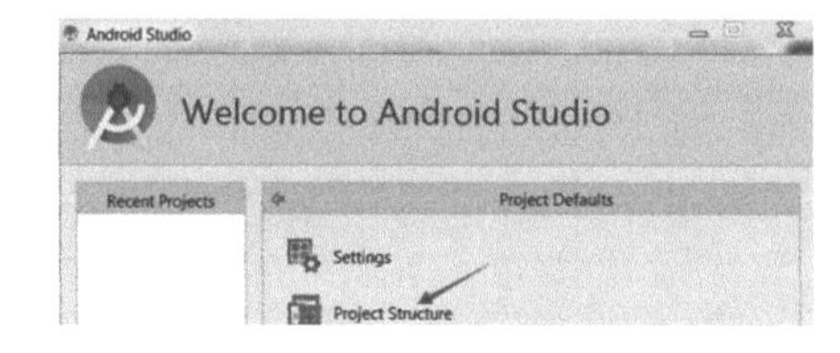

图 1.18　选择“Project Structure”选项

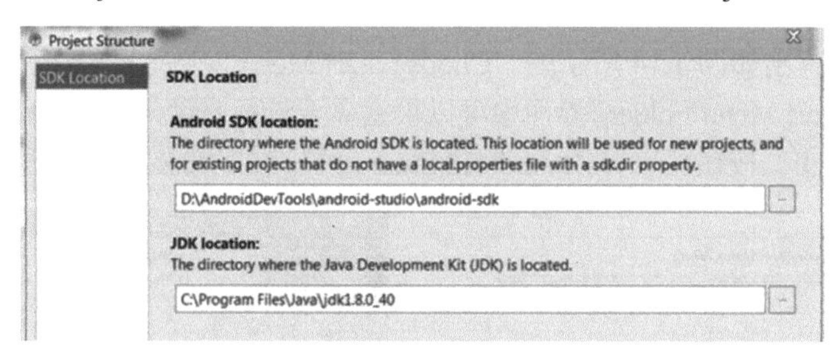

图 1.19　“Project Structure”界面

1.4.3　创建第一个 Android 应用程序

1．创建 HelloWorld 应用程序

在 Android Studio 启动界面中选择“Start a new Android Studio project”，如图 1.20 所示。

图 1.20　选项“Start a new Android Studio project”

填写应用程序名称和项目位置，如图 1.21 所示。

图 1.21　填写应用程序名称和项目位置

选择 Android 设备，如图 1.22 所示。

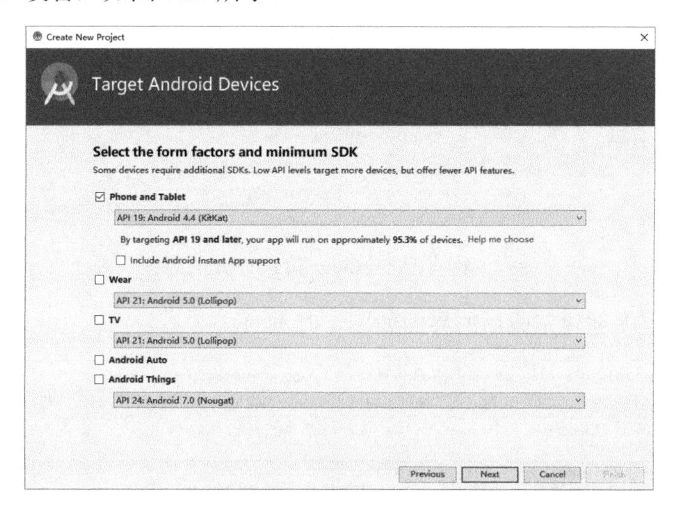

图 1.22　选择 Android 设备

选择“Empty Activity”，如图 1.23 所示。

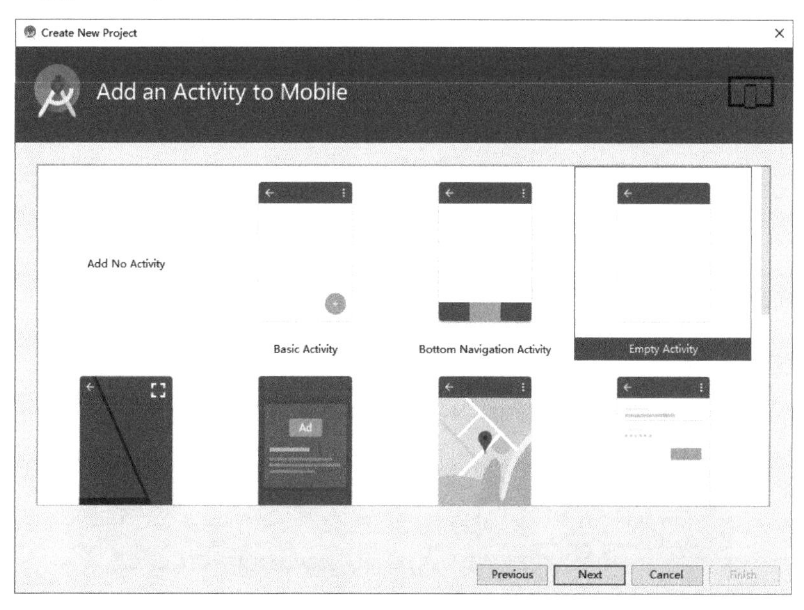

图 1.23　选择“Empty Activity”

输入 Activity 和 Layout 的名称，如图 1.24 所示，单击“Finish”按钮。

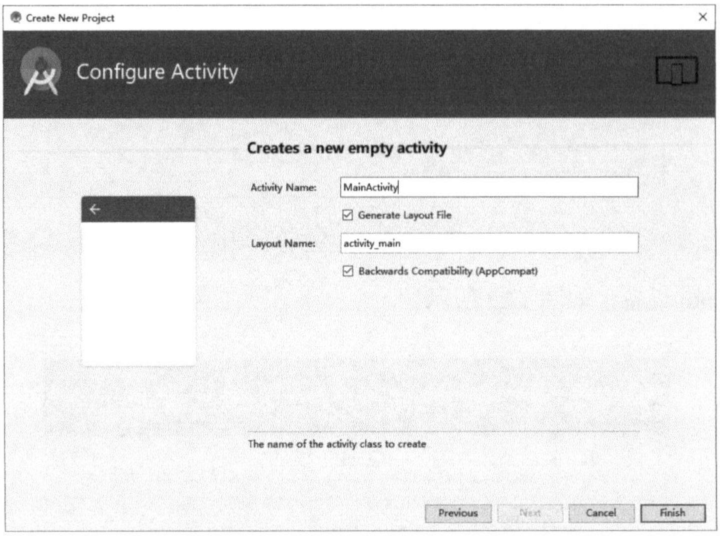

图 1.24　输入 Activity 和 Layout 的名称

Android 应用程序创建完成后的界面如图 1.25 所示。

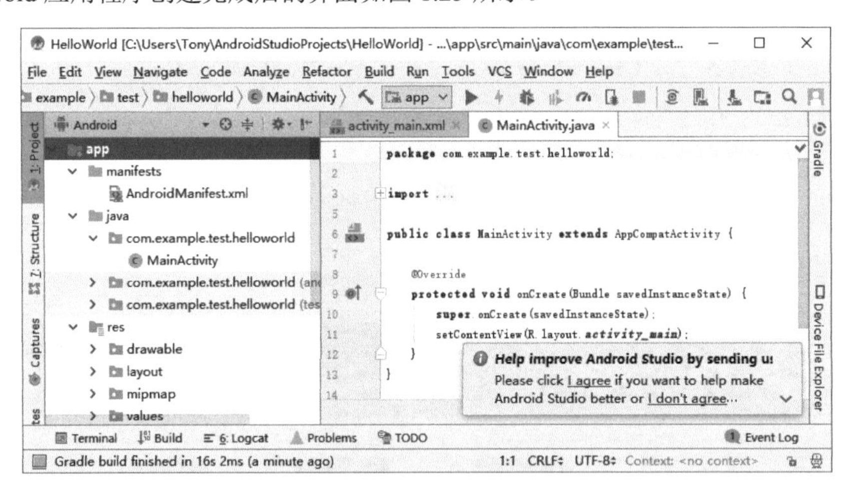

图 1.25　Android 应用程序创建完成后的界面

第一个简单的 Android 应用程序就创建完成了。

2. 创建虚拟机设备

单击图 1.26 中的“app”按钮即可运行创建的 Android 应用程序。

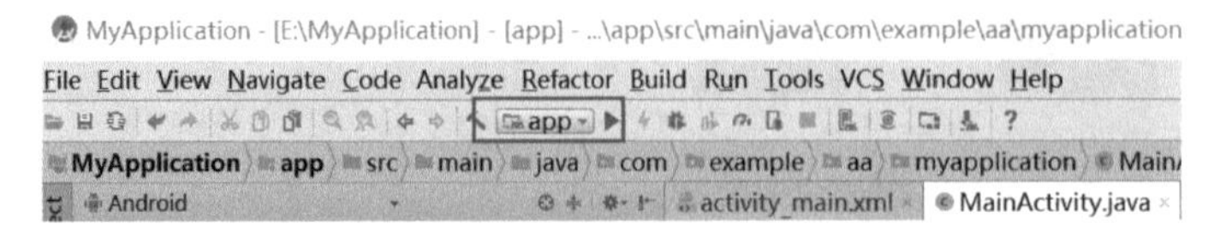

图 1.26　“app”按钮

运行 Android 应用程序后选择硬件，如图 1.27 所示。

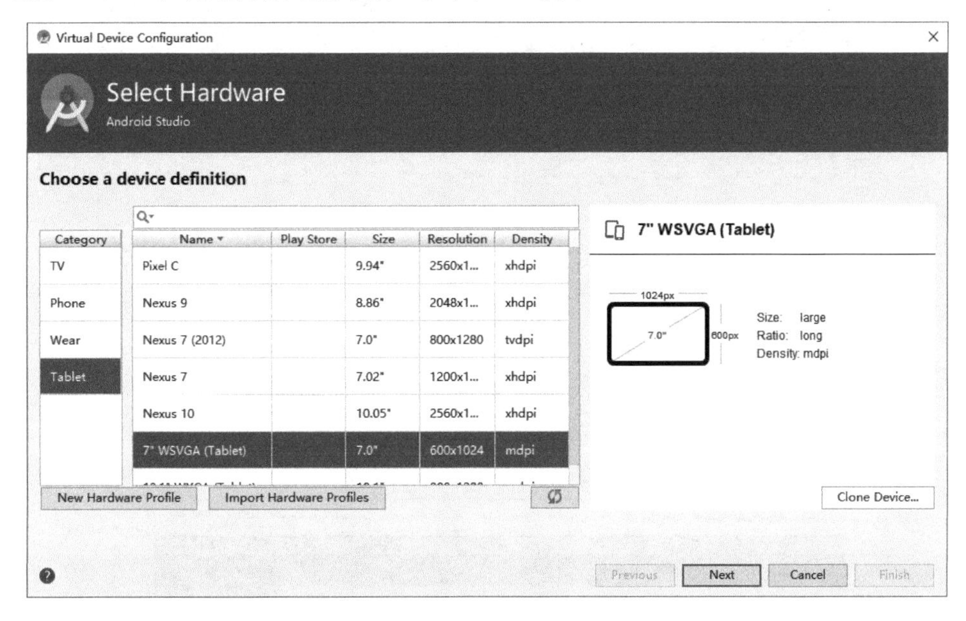

图 1.27　选择硬件

接着选择 Android 系统镜像，如图 1.28 所示。

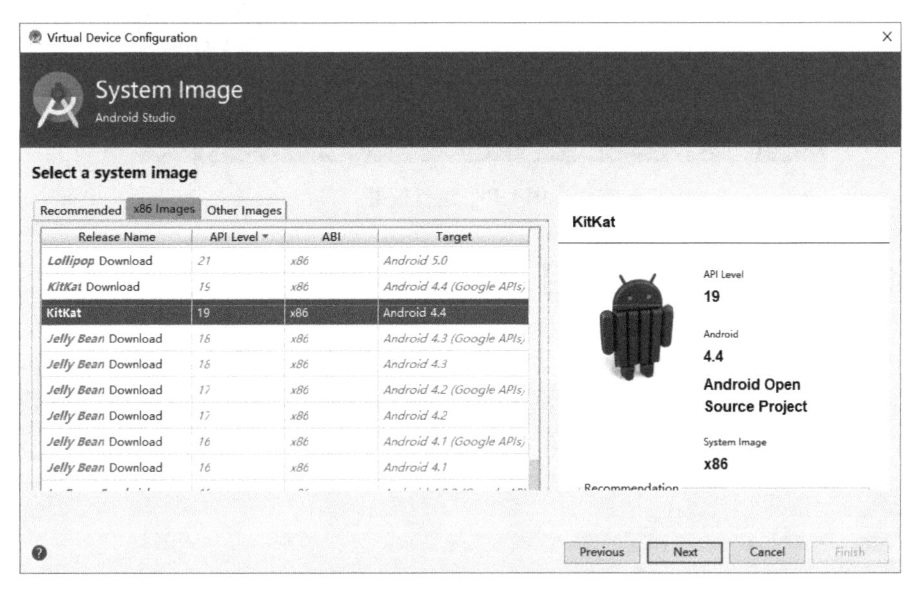

图 1.28　选择 Android 系统镜像

配置 Android 虚拟设备，如图 1.29 所示。

运行效果如图 1.30 所示。

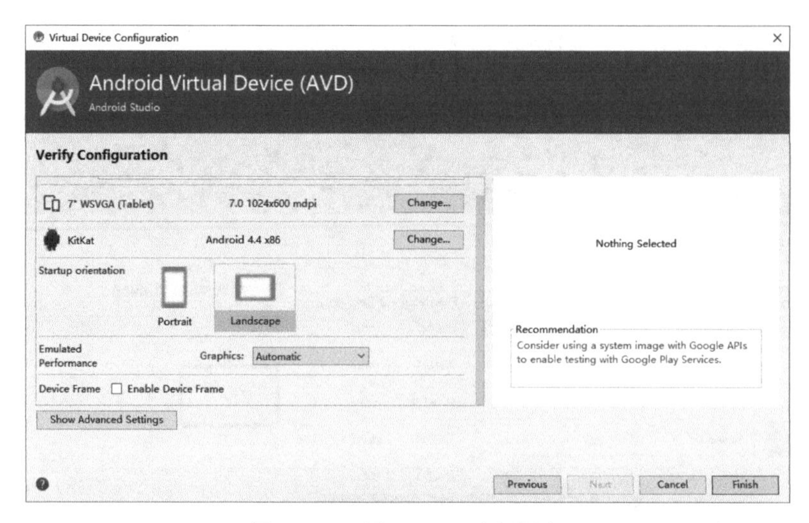

图 1.29　配置 Android 虚拟设备

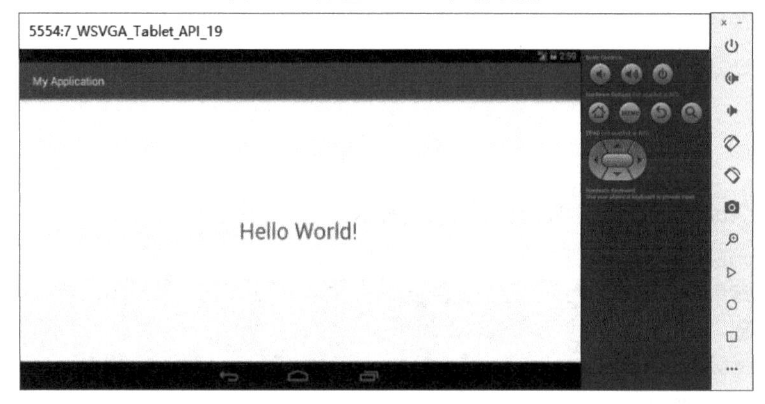

图 1.30　运行效果

1.4.4　导入 Android 项目

在 Android studio 开发环境中，单击“File→Open File or Project”，在弹出的“Open File or Project”界面中选择 Android 项目所在的路径，如图 1.31 所示。单击“OK”按钮后即可导入 Android 项目，导入过程界面如图 1.32 所示。

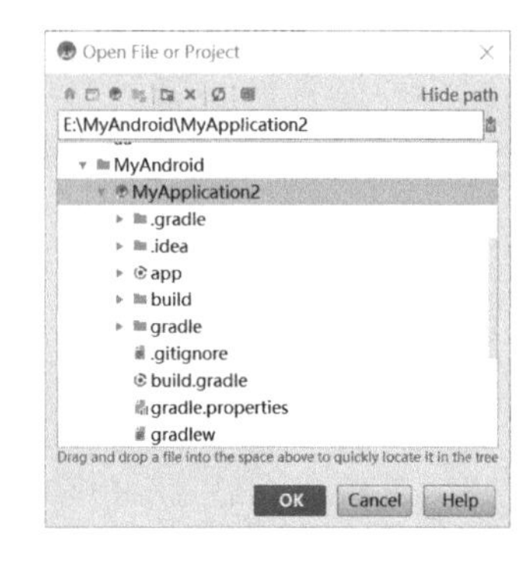

图 1.31　选择 Android 项目所在的路径

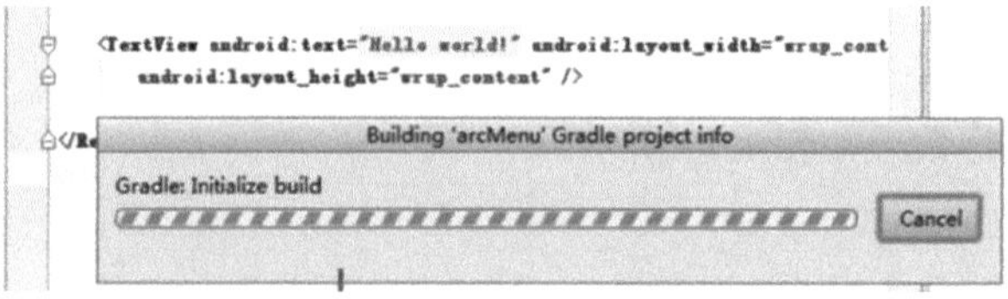

图 1.32　导入过程界面

Android 项目导入成功后的界面如图 1.33 所示。

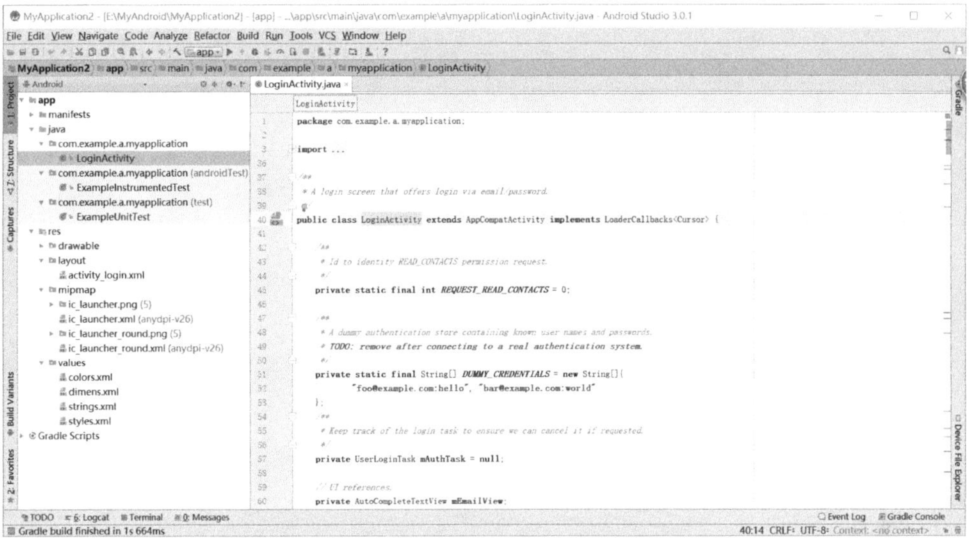

图 1.33　Android 项目导入成功后的界面

1.5　任务验证

通过本任务的学习，读者可以在计算机中搭建 Android Studio 开发环境，并创建第一个 Android 应用程序，还可以导入本书配套资料提供的例程。

1.6　开发小结

本任务简要介绍了 Android 系统架构和开发框架、JDK 的安装与配置、Android Studio 的安装和配置、Android 应用程序的创建、虚拟设备的创建，以及 Android 项目的导入。

1.7　思考与拓展

（1）Android 系统架构一共有几层？

（2）请尝试创建一个 Android 应用程序。

任务2 Java开发基础

本任务主要介绍 Java 开发基础。通过本任务的学习，读者可以了解 Java 的基础语法、对象与类，熟练掌握 Java 开发环境的安装和配置，以及 Java 程序运行和调试。

2.1 开发场景：开发 Android 应用程序需要的程序设计语言

任务 1 介绍了 Android 开发环境的搭建，以及创建 Android 应用程序的方法。Java 是目前 Android 应用开发的主流语言，本任务重点介绍 Java 开发基础。Java 开发框架如图 2.1 所示。

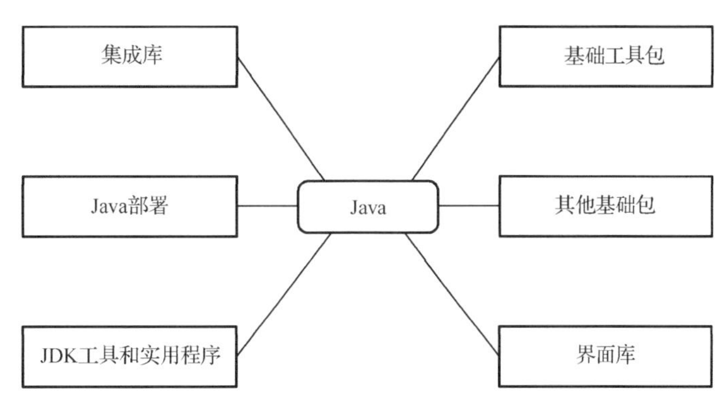

图 2.1 Java 开发框架

2.2 开发目标

（1）知识目标：熟悉 Java 开发环境、Java 基础语法，以及 Java 的对象与类。

（2）技能目标：熟悉 Java 开发环境的搭建，以及 Java 程序的运行及调试。

（3）任务目标：掌握 Java 的基础语法、Java 的对象与类，熟练掌握 Java 开发环境的搭建，熟悉 Java 程序的运行及调试。

2.3 原理学习：Java 开发基础

2.3.1 Java 简介

Java 是面向对象的编程语言，吸收了 C++语言的优点。作为静态面向对象编程语言的代表，可以使用 Java 语言编写桌面应用程序、Web 应用程序、分布式系统应用程序和嵌入式系统应用程序等。

2.3.2 Java 的基本语法

一个 Java 程序可以认为是一系列对象的集合，这些对象通过调用彼此的方法来协同工作。

（1）基本语法。在编写 Java 程序时，应注意以下几点：

- 大小写敏感：Java 程序是大小写敏感的。
- 类名：类名的首字母应该大写，如 MyFirstJavaClass。
- 方法名：所有的方法名都应该以小写字母开头。如果方法名中包含若干个单词，则后面的每个单词首字母大写。
- 源文件名：源文件名必须和类名相同。当保存源文件时，应该使用类名作为源文件名来保存，源文件名的后缀为“.java”。
- 主方法入口：所有的 Java 程序都是由 public static void main(String args[])方法开始执行的。

（2）Java 的标识符。Java 程序所有的组成部分都需要名字。类名、变量名以及方法名都被称为标识符，标识符有以下特点：

- 所有的标识符都应该以字母（A～Z 或者 a～z）、“$”或者“_”开始。
- 首字符之后可以是任何字符的组合。
- 关键字不能作为标识符。
- 标识符大小写敏感。

例如，key、student、_value、_value_1 是合法的标识符，123abc、-tree 是非法标识符。

（3）Java 的修饰符。Java 可以使用修饰符来修饰类中方法和属性。主要有两类修饰符：

- 可访问修饰符，如 default、public、protected、private。
- 不可访问修饰符，如 final、abstract、strictfp。

（4）Java 的变量。Java 主要有局部变量、类变量（静态变量）和成员变量（非静态变量）等。

（5）Java 的数组。数组可以保存多个同类型的变量。

（6）Java 的枚举。枚举中的变量只能是预先设定的值，使用枚举可以减少代码中的错误。例如，为咖啡店设计一个程序，它将咖啡杯限制为小杯、中杯、大杯，这就意味着不允许顾客点这三种规格外的咖啡。

```
class Coffee{
    enum CoffeeSize{ SMALL, MEDUIM, LARGE }
    CoffeeSize size;
}
public class CoffeeTest {
    public static void main(String args[])
    {
        Coffee cof = new Coffee ();
        cof.size = Coffee. CoffeeSize.MEDUIM ;
    }
}
```

（7）Java 的注释。Java 支持单行和多行注释，单行注释的格式为“//注释内容”，多行注释的格式为“/*注释内容*/”。

2.3.3　Java 的对象与类

（1）对象：对象是类的一个实例，有状态和行为。例如，一只猫是一个对象，其状态有颜色、名字、品种，行为有摇尾巴、叫、吃等。

（2）类：类是一个模板，用于描述对象的行为和状态。

2.4　开发实践：Java 开发环境及程序的运行调试

2.4.1　Java 程序的运行机制

Java 程序由 Java 语句、Java 类文件格式、Java 虚拟机和 Java 应用程序接口构成，在编辑、运行 Java 程序时，需要同时涉及这四种方面。

Java 程序的运行机制是：使用文字编辑软件（如记事本）或集成开发环境（如 Eclipse、MyEclipse）在 Java 源文件中定义不同的类，通过调用类中的方法来访问资源系统，把源文件编译生成的二进制字节码存储在 class 文件中，通过与操作系统相对应的 Java 虚拟机来运行 class 文件，在执行编译生成的字节码时会通过 class 文件中的方法来实现 Java 应用程序接口（API）的调用。Java 程序运行机制如图 2.2 所示。

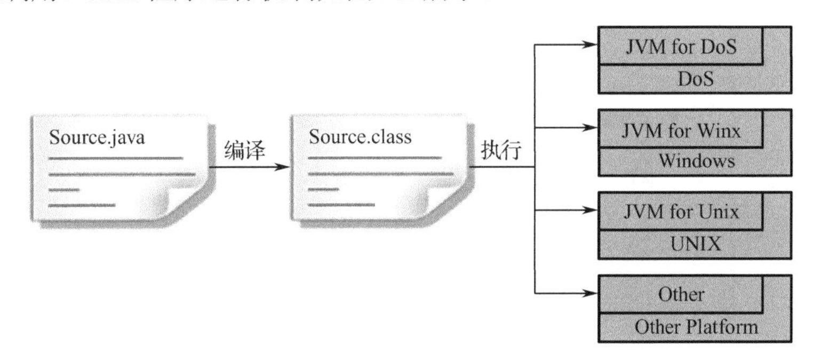

图 2.2　Java 程序运行机制

2.4.2 基于 Android Studio 开发环境开发 Java 程序

本任务介绍 Java 开发基础的目的是更好地开发 Android 应用程序，所以对 Java 的学习也都是在 Android Studio 开发环境中进行的。Android 项目的创建过程如下：

（1）Android 项目的创建界面如图 2.3 所示。

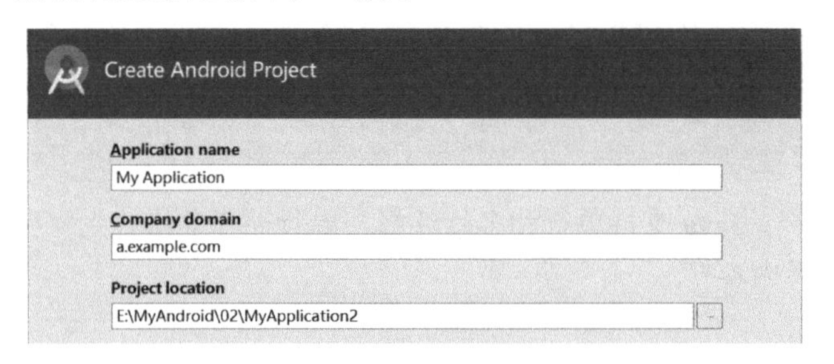

图 2.3 Android 项目的创建界面

（2）新建一个 Module。右键单击 Android 项目中的目录“app”，在弹出的快捷菜单中选择“New→Module”即可新建一个 Module，如图 2.4 所示。

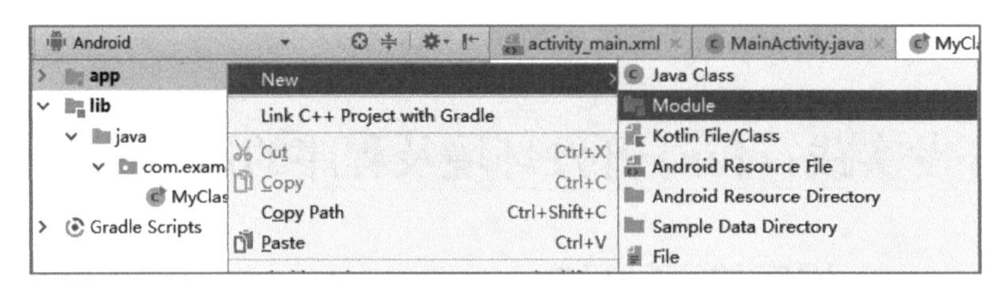

图 2.4 在 Android 项目中创建 Module

（3）选择 Java Library。在如图 2.5 所示的“New Module”界面中选择“Java Library”，这时会创建一个目录“Java Library”，在该目录下面可以编写和运行 Java 程序。

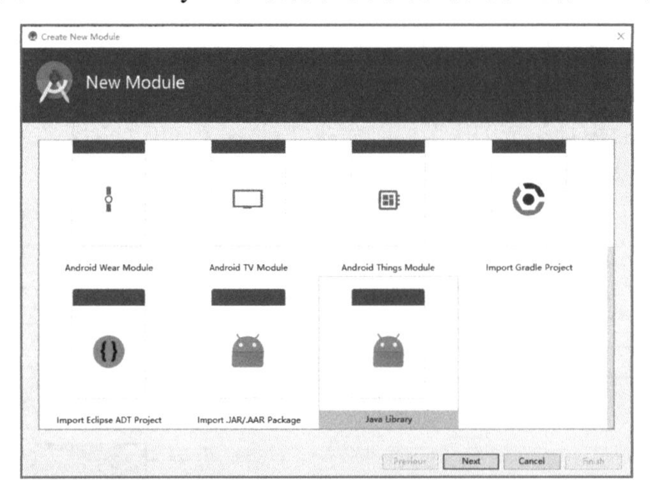

图 2.5 “New Module”界面

（4）创建第一个 Java 类。在新的目录中创建第一个 Java 类，Java 类的代码如图 2.6 所示。

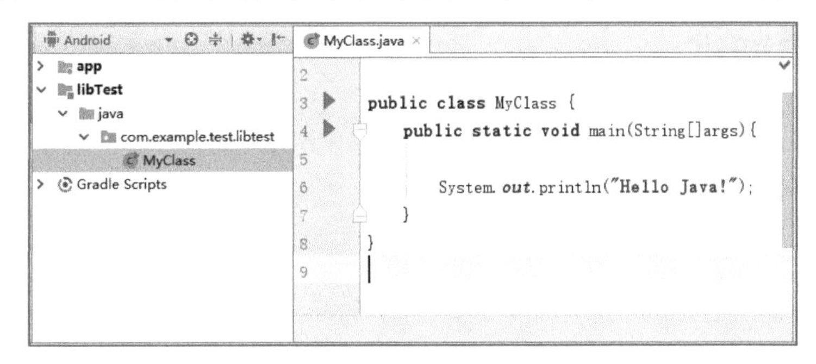

图 2.6　Java 类的代码

在项目的目录中选择类文件后单击鼠标右键，在弹出的快捷菜单中选择“Run ‘MyClass.main()’”即可运行 Java 程序，如图 2.7 所示。

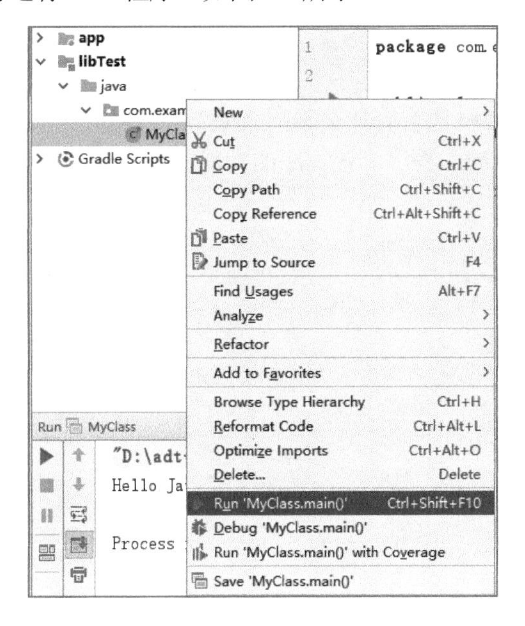

图 2.7　运行 Java 程序

运行结果如图 2.8 所示。

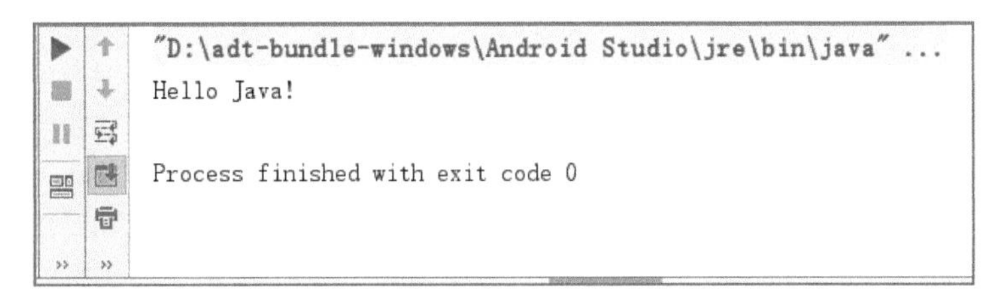

图 2.8　运行结果

2.5 任务验证

通过本任务的学习，读者可以掌握 Java 的基本语法，理解 Java 的对象和类，能够搭建 Java 开发环境并运行 Java 程序。

2.6 开发小结

本任务简要介绍了 Java 的基本语法、对象与类，以及 Java 程序的运行机制，并给出了一个基于 Android Studio 开发环境运行 Java 程序的实例。

2.7 思考与拓展

（1）Java 语言与 C、C++语言有什么不同之处？
（2）请尝试创建一个 Java 程序，该程序的功能是输出自己的名字。

任务3

海拔高度数据的定义与转换

本任务通过海拔高度数据的定义与转换来介绍 Java 的数据类型。通过本任务的学习，读者可以掌握 Java 的运算符、关键字以及程序结构，实现海拔高度数据的定义与转换。

3.1　开发场景：如何用 Java 定义海拔高度数据

我们的日常生活离不开地理数据，海拔高度是一种常见的地理数据。应该如何用 Java 来定义海拔高度数据呢？本任务通过介绍 Java 的基本知识来解答这个问题。

3.2　开发目标

（1）知识目标：熟悉 Java 的数据类型，掌握 Java 的修饰符、运算符、关键字、程序结构。

（2）技能目标：能进行 Java 的数据类型及其转换，实现海拔高度采集与转换程序的开发及运行。

（3）任务目标：实现海拔高度数据的定义与转换。

3.3　原理学习：Java 的数据类型、运算符、关键字、程序结构

3.3.1　Java 的数据类型

Java 有 4 种基本数据类型，可以用来存储数值、字符和布尔值。

（1）整型。整型变量用来存储整数数值，即没有小数部分的数值。整型数据有 3 种表示形式：十进制、八进制和十六进制。根据整型数据所占内存空间的不同可分为 4 种类型，如表 3.1 所示。

表3.1 整型数据的分类（根据所占内存空间的不同分类）

数据类型	所占内存空间	数据类型	所占内存空间
byte	8 bit	short	16 bit
int	32 bit	long	64 bit

（2）浮点型。浮点型数据可分为两类，如表3.2所示。

表3.2 浮点型数据的分类（根据所占内存空间的不同分类）

数据类型	内存空间
float	32 bit
double	64 bit

（3）字符型。字符型变量用于存储单个字符，占用16 bit（2 B）的内存空间。定义字符型变量时要使用关键字char，变量名称要使用单引号。

（4）布尔型。布尔型变量又称为逻辑型变量，是通过关键字 boolean 来定义的。布尔型变量只有“真”和“假”两个取值，分别代表布尔逻辑中的“真”和“假”。布尔型变量在流程控制中常常作为判断条件。

3.3.2 Java的运算符及关键字

1. Java的运算符

运算符主要用于数学函数、一些类型的赋值语句和逻辑比较。

（1）赋值运算符。赋值运算符是一个二元运算符，符号为“=”，其功能是将等号右边的操作数的值赋给等号左边的操作数。例如：

```
int a = 100;
```

（2）算术运算符。算术运算符如表3.3所示。

表3.3 算术运算符

运算符	说明	运算符	说明
+	加	-	减
*	乘	/	除
%	取余数	—	—

（3）自增运算符和自减运算符。自增运算符和自减运算符是单目运算符，可以放在变量之前，也可以放在变量之后。变量一般为整型变量。自增运算符或自减运算符的作用是使变量的值增1或减1。放在变量之前的自增运算符或自减运算符，先将变量的值加1或减1，再使用该变量参与表达式的运算；放在变量之后的自增运算符或自减运算符，先使用变量参与表达式的运算，再将该变量的值加1或减1。例如：

```
//假设 a=8
b=++a;          //先将 a 的值加 1，然后赋值给 b，此时 a 的值为 9，b 的值为 9
b=a++;          //先将 a 的值赋值给 b，再将 a 的值变为 9，此时 a 的值为 9，b 的值为 8
```

（4）比较运算符。比较运算符为二元运算符，用于数值的比较。比较运算符的运算结果是布尔型数据。当运算符对应的关系成立时，运算的结果为真，否则为假。比较运算符共有 6 个，通常作为判断的依据用于条件语句中，如表 3.4 所示。

表 3.4　比较运算符

运 算 符	说　明	运 算 符	说　明
>	比较左方是否大于右方	<	比较左方是否小于右方
==	比较左方是否等于右方	>=	比较左方是否大于或等于右方
<=	比较左方是否小于或等于右方	!=	比较左方是否不等于右方

（5）逻辑运算符。逻辑运算符如表 3.5 所示。

表 3.5　逻辑运算符

运 算 符	说　明	运 算 符	说　明
&	逻辑与	|	逻辑或
!	逻辑非	&&	短路与
||	短路或	^	逻辑异或

（6）位运算符。位运算符如表 3.6 所示。

表 3.6　位运算符

运 算 符	说　明
&	如果对应位都是 1，则结果为 1，否则为 0
|	如果对应位都是 0，则结果为 0，否则为 1
^	如果对应位的值相同，则结果为 0，否则为 1
~	按位取反运算符，翻转操作数的每一位，即 0 变成 1，1 变成 0
<<	按位左移运算符，左边操作数按位左移右边操作数指定的位数
>>	按位右移运算符，左边操作数按位右移右边操作数指定的位数
>>>	按位右移补零操作符，左边操作数的值按右边操作数指定的位数右移，移动产生的空位以零填充

（7）条件运算符。条件运算符是三目运算符，该运算符有 3 个操作数，如果布尔表达式的值为真，则返回“表达式 1”的值，否则返回“表达式 2”的值。条件运算符的格式如下：

```
布尔表达式? 表达式 1: 表达式 2
```

2．Java 的关键字

Java 的关键字用来表示一种数据类型或者表示程序的结构等，关键字不能用于变量名、方法名、类名、包名和参数等。Java 的关键字如表 3.7 所示。

表 3.7 Java 的关键字

关键字	说明	关键字	说明
abstract	抽象类或方法	assert	用来查找内部程序错误
break	跳出一个 switch 或循环	byte	8 位整型
case	switch 的一个分支	catch	捕获 try 代码块中抛出的异常信息，并执行 catch 代码块中的代码
class	定义一个类的类型	continue	不再执行循环体中 continue 语句之后的代码，直接进行下一次循环
default	switch 的默认语句	do	do/while 循环最前面的语句
double	双精度浮点数类型	else	if 语句的 else 语句
enum	枚举类型	extends	定义一个类的父类
final	一个常量或不能覆写的类或方法	finally	try 代码块中总会执行的部分
float	单精度浮点数类型	for	循环类型
if	条件语句	implements	定义一个类实现的接口
import	导入一个包	instanceof	测试一个对象是否是某个类的实例
int	32 位整型	interface	接口，一种抽象类型，仅有方法和常量的定义
long	64 位长整型	native	由宿主系统实现的一个方法
new	分配新的类实例	null	一个空引用
package	包含类的一个包	private	表示私有字段或者方法等，只能从类内部访问
protected	表示保护类型字段	public	表示共有属性或者方法
return	从一个方法中返回	short	16 位整类
static	表示特性是某个类特有的而不属于这个类的对象	strictfp	对浮点数计算使用严格的规则
super	超类对象或构造函数	switch	选择语句
synchronized	对线程而言是原子的方法或代码块	this	当前类的一个方法或构造函数的隐含参数
throw	抛出一个异常	throws	一个方法可能抛出的异常
transient	表示非永久性的数据	try	捕获异常的代码块
void	表示方法不返回任何值	while	一种循环
volatile	表示字段可能会被多个线程同时访问，而不进行同步	—	—

3.3.3 Java 的程序结构

1. 顺序结构

如果代码里没有流程控制，那么程序就会按照从上到下的顺序执行，直到程序结束，例如：

```
public class Demo{
    public static void main(String[] args){
        //顺序结构：自上而下一行一行地顺序执行；
```

```
            System.out.println("第一行 1");
            System.out.println("第二行 2");
            System.out.println("第三行 3");
            System.out.println("第四行 4");
        }
}
```

2. 分支结构

如果程序具有多个条件，则需要通过条件判断来决定程序的具体工作。通过判断条件来做选择的语句称为选择语句或分支语句。选择语句有 if 和 switch 两种。

（1）if 语句。if 语句使用布尔型表达式或布尔型值作为选择条件，有以下三种结构形式：

① if 结构。格式如下：

```
if(布尔型表达式){
    条件执行体
}
```

if 后面跟的{}表示一个整体的代码块，称为条件执行体。当条件为真时，执行条件执行体。

```
//判断 8 大于 5 吗?如果大于则输出 8 大于 5
if(8 > 5){
    System.out.println("8 大于 5");
}
```

② if…else 结构。格式如下：

```
if(布尔型表达式){
    条件执行体 A
}else{
    条件执行体 B
}
```

例如：

```
//求最大值和最小值
//if…else 结构的例子
int x = 38;
int y = 92;
if(x > y){
    System.out.println("x 大于 y");
}else{
    System.out.println("x 小于 y");
}
//相当于三元运算符
String result = x > y ? "x 大于 y" :"x 小于 y";
System.out.println(result);
```

③ if…else if…else 结构。格式如下：

```
if(布尔型表达式 A){
    条件执行体 A
}else if(布尔型表达式 B){
    条件执行体 B
}else{
    条件执行体 C
}
```

例如：

```
int season = 2;
if(season == 1){
    System.out.println("春季");
}else if(season == 2){
    System.out.println("夏季");
}else if(season == 3){
    System.out.println("秋季");
}else if(season == 4){
    System.out.println("冬季");
}else{
    System.out.println("输入有误!");
}
```

（2）switch 语句。格式如下：

```
switch(整型表达式){
    case A:
        当表达式的结果等于 A 时，执行此语句;
        break;
    case B:
        当表达式的结果等于 B 时，执行此语句;
        break;
    default:
        当以上条件都不满足时，执行此语句;
}
```

注意：case 后面的表达式必须是常量。例如：

```
int season = 3;
switch(season ){
    case 1:
        System.out.println("春季");
        break;
    case 2:
        System.out.println("夏季");
        break;
    case 3:
        System.out.println("秋季");
        break;
```

```
        case 4:
            System.out.println("冬季");
            break;
        default:
            System.out.println("输入有误!");
}
```

3．循环结构

（1）while 循环。格式如下：

```
while(布尔型表达式){
    循环体
}
```

while 循环的特点是：先判断表达式，若为 true 则执行循环体，否则跳过循环体。例如：

```
//计算 10 以内的正整数之和
int num = 1;
int result = 0;
while(num <= 10){
    result += num;
    num ++;
}
System.out.println("10 以内的正整数之和为: " + result);
```

（2）do…while 循环。格式如下：

```
do{
    循环体语句
}while(布尔型表达式);
```

do…while 循环的特点是：先执行一次循环体，再判断表达式，若为真则就执行循环体，否则就跳过循环体。也就是说，do…while 是先执行后判断，即使判断条件为假，也会至少循环一次。

（3）for 循环。格式如下：

```
for(初始化语句;布尔型表达式; 循环后操作语句){
    循环体语句
}
```

初始化语句：表示对循环进行初始化，只在循环开始时执行一次，定义一个变量并赋值。

布尔型表达式：当布尔型表达式为 false 时，循环终止；当布尔型表达式为 true 时，才会执行循环体语句。

循环后操作语句：每次循环后都会调用该语句，一般该语句是递增或递减操作。

例如：

```
//计算 10 以内的正整数之和
int result = 0;
for(int i= 1; i<= 10; i++){
```

```
        result += i;
    }
    System.out.println("10 以内的正整数之和为: " + result);
```

（4）嵌套循环。若外循环的循环次数是 m 次，在每一次外循环时内循环的循环次数是 n 次，则总的内循环的循环次数是 mn 次。使用嵌套循环时应注意以下 3 点：

① 使用循环嵌套时，内循环和外循环的循环控制变量不能相同。

② 循环嵌套结构的书写最好采用“右缩进”格式，以体现循环层次的关系。

③ 尽量避免太多和太深的循环嵌套结构。

例如：

```
//打印三角形：外层控制行数，内层控制星号的个数
    for (int i = 1; i <=5; i++) {
        for(int j=1;j<=i;j++){
            System.out.print("*");
        }
        System.out.println();
}
```

4．循环结构控制语句

（1）break 语句。break 语句用于终止当前所在的循环。注意：break 之后的语句不执行，所以不能编写，否则会在编译时报错。例如：

```
//输出 1 到 10
for(int i = 1; i <= 10; i ++){
    //如果输出 7，则终止循环
    if(i == 7){
        break;                  //跳出循环或循环终止
    }
    System.out.println(i);
}
System.out.println("输出完成！");
//输出结果
1
2
3
4
5
6
输出完成！
```

（2）continue 语句。continue 语句用于跳过当前的循环，进入下一次循环操作。例如：

```
//输出 1 到 10
for(int i = 1;i <= 10;i ++){
    //如果输出 8，则跳过当前的循环
    if(i == 8){
        continue;                       //跳过当前的循环，进入下一次循环
```

```
            System.out.println("------");          //当前的循环体之后的语句不再执行
        }
        System.out.println(i);
}
System.out.println("输出完成！");
//输出结果
1
2
3
4
5
6
7
9
10
输出完成！
```

（3）return 语句。return 语句用于结束循环所在的方法，循环所在的方法结束了，循环结构也就结束了。由于 break、continue、return 后面的语句不会执行，所以它们后面不能再跟任何语句，否则会在编译时出错。break 和 return 语句都能结束当前的循环，但如果循环结构之后的语句还得执行，那么就应该使用 break 语句。例如：

```
for(int i = 1;i <= 10;i ++){
        if(i == 6){
            return;                              //结束循环所在的方法
        }
        System.out.println(i);
}
System.out.println("输出完成！");               //这条语句不会执行
//输出结果
1
2
3
4
5
```

3.4　开发实践：海拔高度数据的定义与转换

3.4.1　开发设计

本任务假设海拔高度数据是从第三方数据接口获取的。本任务中的程序先将获取到的海拔高度数据返回到 String 类型数组中，然后将海拔高度数据保存到整型数组中，接着将 String 类型数据转换成整型数据并将其读取出来，最后找出其中最大的数据并打印出来。主程序执行流程如图 3.1 所示。

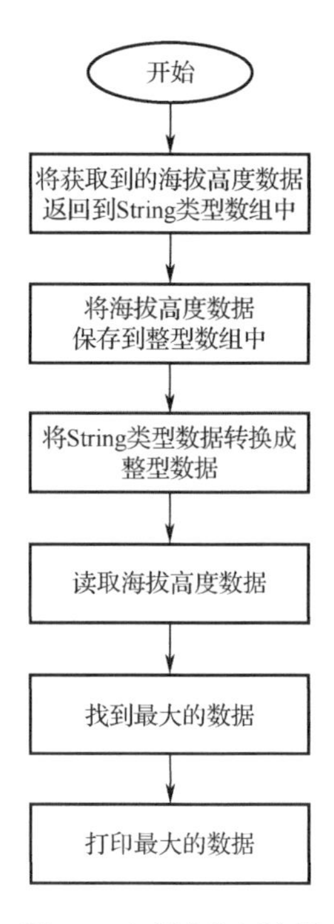

图 3.1　主程序执行流程

3.4.2　功能实现

1．模拟海拔高度数据模块

```
public class DataGenerater {
    /*将海拔高度数据保存在 String 类型数组中*/
    private static String[] Elevations = {"8848","120","301","4003","560","1001","2008","3139","561"};
    public static String[] getElevation() {
        return Elevations;
    }
}
```

2．主函数模块

```
public static void main(String[] args) {
    //获取海拔高度数据并返回到 String 类型数组中
    String[] elevationsStr = DataGenerater.getElevation();
    //将获取到的海拔高度数据保存到整型数组中
    int size = elevationsStr.length;
    int[] elevations = new int[size];
    for (int i = 0; i < size; i++) {
        //将 String 类型数据转换成整型数据
        int elevation = Integer.valueOf(elevationsStr[i]);
        elevations[i] = elevation;
    }
```

```
        //读取海拔高度数据
        for (int i = 0; i < size; i++) {
            System.out.println("第"+i+"个海拔高度："+elevations[i]);
        }
        //查找最大的数据
        int high = 0;
        for (int i = 0; i < size; i++) {
            if(high < elevations[i]) {
                high = elevations[i];
            }
        }
        //输出最大的数据
        System.out.println("当前最高的海拔高度："+high);
    }
```

3.5　任务验证

首先将本任务的例程导入 Android Studio 开发环境并定位到 TestClass 文件，然后运行该文件，如图 3.2 所示。

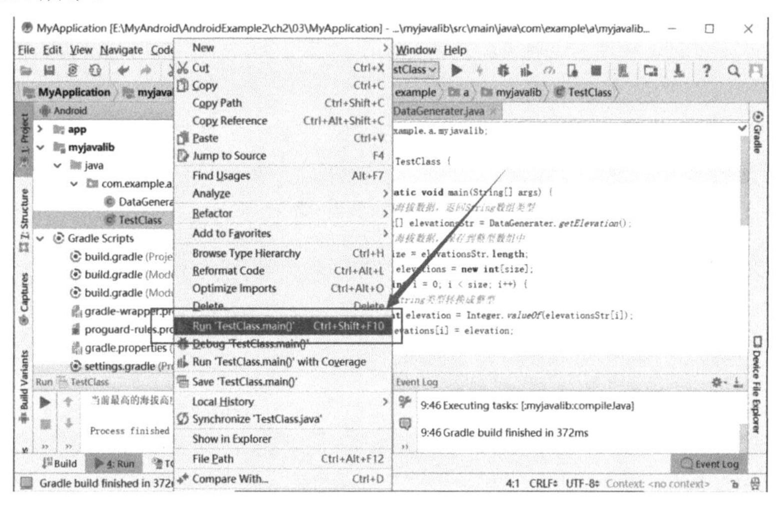

图 3.2　运行 TestClass 文件

输出的海拔高度数据如图 3.3 所示。

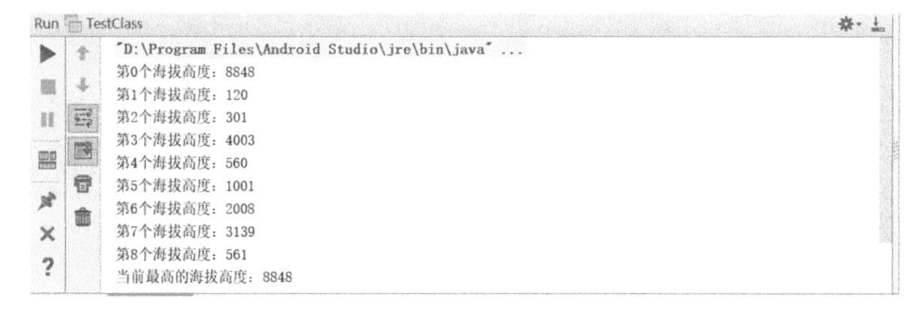

图 3.3　输出的海拔高度数据

3.6 开发小结

本任务主要介绍了 Java 的数据类型、运算符、关键字及程序结构，并实现了海拔高度数据的定义与转换。

3.7 思考与拓展

（1）Java 运算符中的“==”与“=”有何区别？

（2）如何把浮点型数据转换成 String 类型数据？

（3）请尝试修改本任务的程序，当海拔高度大于 1000 m 时，输出 0，否则输出 1。

任务 4

温度和湿度数据的记录

本任务在任务 3 的基础上介绍 Java 中的复杂数据类型，包括 Java 数组、Java 常用工具类，通过记录温度和湿度数据来加深读者对 Java 中复杂数据类型的理解。

4.1 开发场景：如何记录温度和湿度数据

天气观测数据是天气预报的基础，这些数据包括温度、湿度、气压、风向、风速、降水、能见度、地温等。本任务介绍如何在 Java 中记录温度和湿度数据。

4.2 开发目标

（1）知识目标：熟悉 Java 中的 Number 类、Math 类、Character 类和 String 类，熟悉 Java 数组、Java 方法。

（2）技能目标：熟悉常用的 Java 类的使用，掌握 Java 数组和 Java IO 流的使用。

（3）任务目标：在 Java 中记录温度和湿度数据。

4.3 原理学习：熟悉 Java 类、数组、方法、IO 流

4.3.1 常用的 Java 类

1. Number 类和 Math 类

Java 为每一个数据类型提供了对应的封装类，这些封装类（如 Integer、Long、Byte、Double、Float、Short）都是抽象类 Number 的子类。

由编译器支持的包装称为装箱。当内置数据类型被当成对象使用时，编译器会把内置数据类型装箱为包装类。编译器也可以把一个对象拆箱为内置数据类型。Number 类属于 java.lang 包。

Java 的 Math 类包含了用于执行基本数学运算的属性和方法，如指数、对数、平方根和

三角函数等。Math 类的方法都被定义为静态形式，可以在主函数中直接调用。表 4.1 列出的是 Number 类和 Math 类的常用方法。

表 4.1　Number 类和 Math 类的常用方法

方 法	描 述	方 法	描 述
xxxValue()	xxx 为某种数据类型，将 Number 对象转换为 xxx 数据类型的值并返回	valueOf()	返回一个 Number 对象指定的内置数据类型
compareTo()	将 Number 对象与参数比较	equals()	判断 Number 对象是否与参数相等
toString()	以字符串形式返回值	parseInt()	将字符串解析为 int 类型
abs()	返回参数的绝对值	ceil()	返回大于或等于（>=）给定参数的最小整数
cos()	求指定 double 类型参数的余弦值	floor()	返回小于或等于（<=）给定参数的最大整数
rint()	返回与参数最接近的整数，返回类型为 double	round()	表示四舍五入，算法为 Math.floor(x+0.5)，即将原来的数字加上 0.5 后再向下取整，所以 Math.round(11.5) 的结果为 12，Math.round(-11.5)的结果为-11
min()	返回两个参数中的最小值	max()	返回两个参数中的最大值
exp()	返回以自然数 e 为底数的参数次方	log()	返回以自然数 e 为底数的参数的对数值
pow()	返回第一个参数的第二个参数次方	sqrt()	求参数的算术平方根
sin()	求指定 double 类型参数的正弦值	—	—

2. Character 类

Java 为 char 类型提供了包装类，即 Character 类。Character 类用于对单个字符进行操作，Character 类在对象中包装一个基本数据类型 char 的值。

Character 类提供了一系列方法来操作字符，可以使用 Character 类的构造函数来创建一个 Character 对象，例如：

```
Character ch = new Character ('a');
```

Java 编译器会自动创建一个 Character 对象。例如，将一个 char 类型的参数传递给需要一个 Character 类型参数的方法时，编译器会自动地将 char 类型参数转换为 Character 对象。这种行为称为装箱，反之称为拆箱。

```
//将原始字符 'n' 装箱到 Character 对象 ch 中
Character ch = 'n';
//将原始字符'm'用 test 方法装箱
//将拆箱的值返回到 c
char c = test('m');
```

前面有反斜杠（\）的字符是转义字符。Java 的转义字符如表 4.2 所示。

表 4.2　Java 的转义字符

转 义 字 符	说 明	转 义 字 符	说 明
\t	在该处插入一个 tab 键	\b	在该处插入一个后退键
\n	在该处换行	\r	在该处插入回车

续表

转义字符	说　明	转义字符	说　明
\f	在该处插入换页符	\'	在该处插入单引号
\"	在该处插入双引号	\\	在该处插入反斜杠

Character 类的常用方法如表 4.3 所示。

表 4.3　Character 类的常用方法

方　法	说　明	方　法	说　明
isLetter()	是否是一个字母	isDigit()	是否是一个数字
isWhitespace()	是否是一个空格	isUpperCase()	是否是大写字母
isLowerCase()	是否是小写字母	toUpperCase()	指定字母的大写形式
toLowerCase()	指定字母的小写形式	toString()	以字符串的形式返回字符，字符串的长度仅为 1

3. String 类

在 Java 中，字符串属于对象，可以使用 Java 提供的 String 类来创建和操作字符串。

（1）创建字符串。创建字符串的方式如下：

```
String Str= "字符串名称";
```

在程序编译过程中，当遇到字符串常量时，编译器会使用字符串的值来创建一个 String 对象。可以使用关键字和构造函数来创建 String 对象，String 类有 11 种构造函数，提供了不同参数来初始化字符串。

（2）字符串长度。在 String 类中，获取字符串长度的方法为 length()。例如，下面的代码执行后，len 变量等于 4。

```
public class StringTest {
    public static void main(String args[]) {
        String str = "Hello";
        int len = str.length();
        System.out.println( "字符串长度 : " + len );
    }
}
```

（3）连接字符串。String 类提供了连接两个字符串的方法 concat()，例如：

```
string1.concat(string2);
```

返回的是 string2 连接 string1 的新字符串。也可以对字符串常量使用 concat()方法，例如：

```
"My name is ".concat("Tom");
```

还可以使用“+”操作符来连接字符串，例如：

```
"Hello," + " Tom " + "!"
```

（4）输出格式化字符串。当需要输出格式化字符串时，可以使用 printf()方法和 format()

方法。String 类使用静态方法 format()，返回一个 String 对象而不是 PrintStream 对象。可以用 String 类的静态方法 format()来创建一个可复用的格式化字符串，而不仅仅是用于一次输出。例如：

```
int a = 5;
float f = 3.65F;
String str = "Hello World!";
System.out.printf("整型变量为" + "%d, 浮点型变量为" + " %f, 字符串变量为" + "%s",   a, f, str);
```

用 format()方法可以改写为：

```
String result = String.format("整型变量为" + "%d, 浮点型变量为" + " %f, 字符串变量为" + "%s",   a, f, str);
```

（5）String 类的常用方法。String 类的常用方法如表 4.4 所示，更多的方法请参考 Java String API 文档。

表 4.4　String 类的常用方法

方　法	说　明
char charAt(int index)	返回指定索引处的 char 值
int compareTo(Object o)	将这个字符串和另一个对象进行比较
int compareTo(String anotherString)	按字典顺序比较两个字符串
int compareToIgnoreCase(String str)	按字典顺序比较两个字符串，不考虑大小写
String concat(String str)	将指定字符串连接到此字符串的结尾
static String copyValueOf(char[] data)	返回指定数组中表示该字符序列的 String
boolean endsWith(String suffix)	测试此字符串是否以指定的后缀结束
boolean equals(Object anObject)	将此字符串与指定的对象进行比较
boolean equalsIgnoreCase(String anotherString)	将此 String 与另一个 String 进行比较，不考虑大小写
byte[] getBytes()	使用平台的默认字符集将此 String 编码为 Byte 序列，并将结果存储到一个新的 Byte 数组中
byte[] getBytes(String charsetName)	使用指定的字符集将此 String 编码为 Byte 序列，并将结果存储到一个新的 Byte 数组中
void getChars(int srcBegin, int srcEnd, char[] dst, int dstBegin)	将字符从此字符串复制到目标字符数组
int length()	返回此字符串的长度
boolean matches(String regex)	告知此字符串是否匹配给定的正则表达式
boolean regionMatches(boolean ignoreCase, int toffset, String other, int ooffset, int len)	测试两个字符串区域是否相等
CharSequence subSequence(int beginIndex, int endIndex)	返回一个新的字符序列，它是此序列的一个子序列
String substring(int beginIndex)	返回一个新的字符串，它是此字符串的一个子字符串
char[] toCharArray()	将此字符串转换为一个新的字符数组
String toLowerCase()	使用语言环境的默认规则将此 String 中的所有字符都转换为小写
String toLowerCase(Locale locale)	使用给定 Locale 的规则将此 String 中的所有字符都转换为小写

4.3.2　Java 数组

Java 用数组来存储固定大小的同类型元素，可以声明一个数组变量（如 numbers[200]）来代替直接声明 200 个独立变量。

1．声明数组变量

先须先声明数组变量，才能在程序中使用数组。例如：

```
DataType[] dataArray;    //首选的方法
```

2．创建数组

Java 使用 new 操作符来创建数组，语法如下：

```
dataArray= new DataType[arraySize];
```

上面语句有两个功能：

（1）使用 DataType[arraySize]创建了一个数组。

（2）把新创建的数组的引用赋值给变量 dataArray。

数组变量的声明和创建可以用一条语句完成，如下所示：

```
DataType[] dataArray= new DataType[arraySize];
```

另外，还可以使用如下的方式创建数组。

```
DataType[] dataArray= {value0, value1, …, value9};
```

数组中的元素是通过索引来访问的，数组的索引从 0 开始。例如，下面的语句首先声明了一个数组变量 testList，接着创建了一个包含 5 个 int 类型元素的数组，并且把它的引用赋值给 testList 变量。例如：

```
public class TestArray {
    public static void main(String[] args) {
        int size = 5;
        int[] testList = new int[size];
        testList [0] = 7;
        testList [1] = 4;
        testList [2] = 3;
        testList [3] = 13;
        testList [4] = 6;
        //计算所有元素的总和
        int total = 0;
        for (int i = 0; i < size; i++) {
            total += testList [i];
        }
        System.out.println("总和为：  " + total);
    }
}
```

3．处理数组

如果数组的元素类型和数组的大小都是确定的，就可以使用 for 循环来处理数组元素。例如：

```
public class TestArray {
    public static void main(String[] args) {
        int[] testList = {7, 2, 3, 8, 5};
        //输出数组的所有元素
        for (int i = 0; i < testList.length; i++) {
            System.out.println(testList [i]);
        }
        //计算所有元素的总和
        int total = 0;
        for (int i = 0; i < testList.length; i++) {
            total += testList [i];
        }
        System.out.println("总和为： " + total)

//输出结果
7
2
3
8
5
总和为：25
```

4．for-each 循环

在不使用下标的情况下遍历数组，可以采用 for-each 循环来处理数组元素。例如：

```
public class TestArray {
    public static void main(String[] args) {
        int[] testList = {7, 2, 3, 8, 5};
        //输出数组的所有元素
        for (int element: testList) {
            System.out.println(element);
        }
    }
}
```

5．二维数组

二维数组的每一个元素都是一个一维数组，如 String str[][] = new String[4][5]。数组的动态初始化如下：

（1）直接为每个一维数组分配空间。

```
DataType arrayName = new DataType [arrayLength1][arrayLength2];
```

DataType 可以是基本数据类型和复合数据类型，arrayLength1 和 arrayLength2 必须为正整数，arrayLength1 为行数，arrayLength2 为列数。例如：

```
int a[][] = new int[4][5];
```

（2）从最高维开始，分别为每个一维数组分配空间。

```
String s[][] = new String[2][];
str[0] = new String[2];
str [1] = new String[3];
str [0][0] = new String("Nice");
str [0][1] = new String("to");
str [1][0] = new String("meet");
str [1][1] = new String("you");
str [1][2] = new String("!");
```

解析：str[0]=new String[2]和 str[1]=new String[3]是为最高维分配的引用空间，也就是能保存数据的最大长度，可为其每个数组元素单独分配空间，如“str[0][0]=new String("Nice");”。

（3）二维数组的引用。二维数组元素的引用方式为 arrayName[index1][index2]。例如：

```
num[2][1];
```

6. Arrays 类

java.util.Arrays 类可以操作数组，它提供的方法具有以下特点：

（1）赋值：通过 fill 方法。

（2）排序：通过 sort 方法，按升序排序。

（3）比较：通过 equals 方法比较数组中元素是否相等。

（4）查找数组元素：通过 binarySearch 方法能对排序好的数组进行二分查找法操作。

Arrays 类的方法如表 4.5 所示。

表 4.5　Arrays 类的方法

方　法	说　明
public static int binarySearch(Object[] a, Object key)	用二分查找法在给定数组中搜索给定值的对象。数组在调用前须排序好，如果查找值包含在数组中，则返回搜索键的索引
public static boolean equals(long[] a, long[] a2)	如果两个指定的 long 类型数组彼此相等，则返回 true
public static void fill(int[] a, int val)	将指定的 int 值分配给 int 类型数组中指定范围的元素
public static void sort(Object[] a)	根据指定对象数组元素的自然顺序进行升序排序

4.3.3　Java 方法

1. Java 方法的命名规则

Java 方法的命名规则如下：

（1）方法名的第一个单词应以小写字母作为开头，后面的单词要首字母大写，不使用连接符，如 addApple。

（2）可以用下画线来区分逻辑组件，如 testPop_emptyStack。

2. Java 方法的定义

定义 Java 方法的格式为：

```
修饰符  返回值类型  方法名（参数类型  参数名）{
    ……
    方法体
    ……
    return 返回值;
}
```

方法包含一个方法头和一个方法体。下面是一个方法的所有部分：

（1）修饰符：定义了该方法的访问类型。

（2）返回值类型：方法返回值的数据类型，如果没有返回值，则类型为 void。

（3）方法名：方法名和参数表共同构成方法签名。

（4）参数类型：当方法被调用时，传递值给参数，方法也可以不包含任何参数。

（5）方法体：方法体包含具体的语句，用于实现该方法的功能。

3. Java 方法的调用

当在程序中调用一个方法时，程序的控制权就交给了被调用的方法，方法被调用后会返回一个值时，因此通常可以将方法的调用当成一个值。例如，求一个最小值的代码如下：

```
int larger = min(10, 30);
```

如果方法被调用后的返回值是 void，则方法的调用一定是一条语句。例如，方法 println 返回 void。下面的调用是个语句：

```
System.out.println("方法调用！");
```

下面的例子演示了如何定义一个方法，以及如何调用它。

```
public class TestMax {
    /*主方法*/
    public static void main(String[] args) {
        int i = 6;
        int j = 4;
        int k = min(i, j);
        System.out.println( i + " 和 " + j + " 比较，最小值是：" + k);
    }
    /*返回两个整数变量中的最小值*/
    public static int min(int num1, int num2) {
        int result;
        if (num1 > num2)
            result = num2;
        else
            result = num1;
        return result;
```

```
        }
    }
//输出结果
6和4比较，最小值是：4
```

4.3.4　Java IO 流

Java 的核心库 java.io 可提供大部分输入、输出操作所需要的类，输入输出（IO）流的类代表了输入源和输出目标，支持多种格式，如基本类型、对象、本地化字符集等。

一个流可以理解为一个数据序列，输入流表示从一个输入源读数据，输出流表示向一个输出目标写数据。

1. IO 流

（1）IO 流。IO 流的类的层次关系如图 4.1 所示。

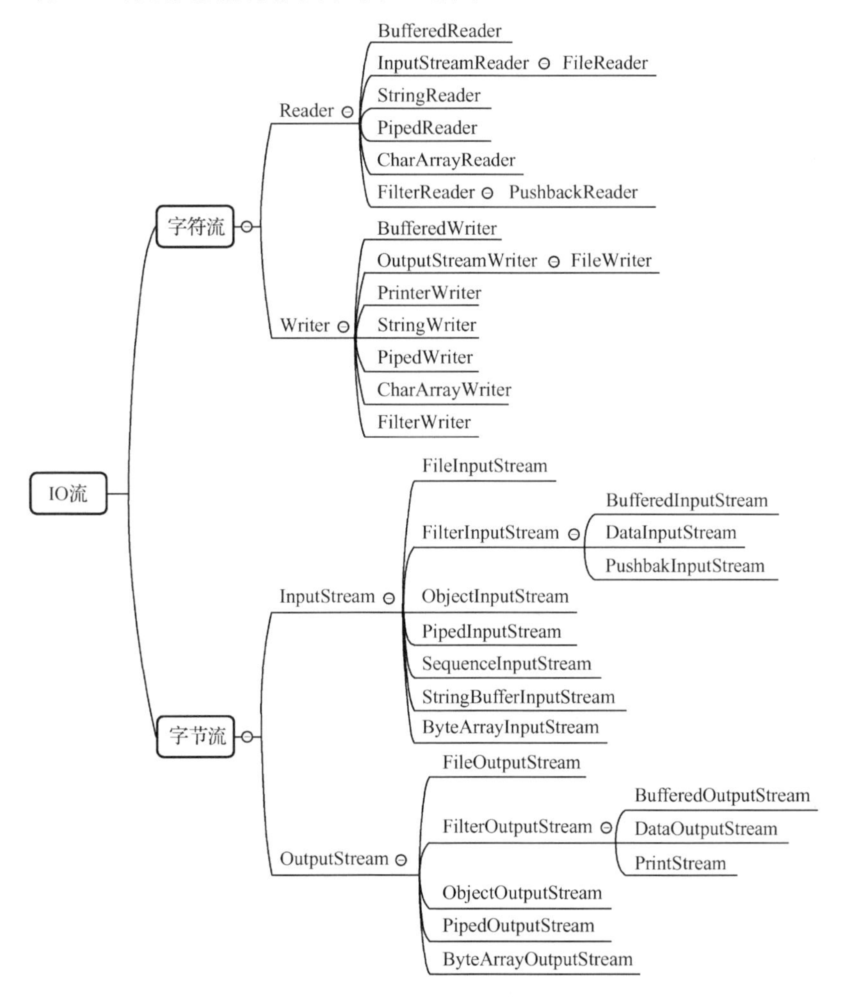

图 4.1　IO 流的类的层次关系

在 java.io 中，用于对文件内容进行操作的主要是字节流和字符流，它们都可以分为输入操作和输出操作。

- InputStream 和 OutputStream：主要用来处理字节或二进制对象，是为字节流设计的。
- Reader 和 Writer：主要用来处理字符或字符串，是为字符流设计的。

（2）字节流转换为字符流。为了便于处理字节流，通常会把字节流转换为字符流。例如，Java 的控制台（Console）输入由 System.in 完成，为了获得一个绑定到控制台的字符流，可以把 System.in 包装在一个 BufferedReader 对象中来创建一个字符流。下面是创建 BufferedReader 对象的方法：

```
BufferedReader bufR = new BufferedReader(new InputStreamReader(System.in));
```

同理，读取文件内容的方法是：

```
InputStream file = new FileInputStream("C:/java/hello");
BufferedReader bufR = new BufferedReader(new InputStreamReader(f));
```

BufferedReader 对象可以实现缓冲功能，java.io.BufferedReader 类和 java.io.BufferedWriter 类都拥有 8192 个字符的缓冲区。当使用 BufferedReader 对象读取文件内容时，会先从文件中读入字符数据并放入缓冲区，然后使用 read()方法从缓冲区读取数据。如果缓冲区数据不足，则再从文件中读取；当使用 BufferedWriter 对象向文件写入数据时，会先把要写入的数据放到缓冲区中。相关的类分析如下：

- InputStreamReader：将字节流输出为字符流，并且为字节流指定字符集。
- BufferedReader：先从输入流中读取文本，再缓存字符，可以高效地读取字符、数组和行。

InputStream 提供的 read 方法可以从输入流中读取 1 B 的数据，常用的函数原型如下：

① public int read() throws IOException {}。该函数原型从输入流中读取数据的下一个字节。返回值为 0～255 的 int 类型字节值。如果已经到达流末尾而没有可用的字节，则返回值-1。在输入数据可用、检测到流末尾或者抛出异常前，read 方法将一直阻塞。

返回：下一个数据字节；如果已到达文件末尾，则返回-1。

② public int read(byte[] b) throws IOException{}。从此输入流中将最多 b.length 个字节的数据读入一个 byte 类型数组中。在某些输入可用之前，read 方法将阻塞。该函数原型覆写了 InputStream 类中的 read 方法；其参数为存储读取数据的缓冲区；其返回值为读入缓冲区的字节总数，如果因为已经到达文件末尾而没有更多的数据，则返回-1。

```
//示例 1：从控制台读取单个字符
public static void main(String[] args) throws IOException{
    BufferedReader BufChar= new BufferedReader(new InputStreamReader(System.in));
    char ch;
    System.out.println("请输入英文字符,输入'e'  字符退出:");
    do {
        ch = (char) BufChar.read();
        System.out.println(ch);
    }while(ch != 'e');
}
//示例 2：从控制台读取多个字符
```

```
public static void main(String[] args) throws IOException{
    BufferedReader BufString= new BufferedReader(new InputStreamReader(System.in));
    String str;
    System.out.println("请输入字符串,输入字符串“quit”退出:");
    do {
        str = BufString.readLine();
        System.out.println(str);
    }while(!str.equals("quit"));
}
//示例 3：使用 InputStream 中的 read 方法读取字节流
public static void main(String[] args) throws IOException{
    InputStream InStr = System.in;
    byte[] buf = new byte[2048];
    int len = 0;
    StringBuffer StrBuf = new StringBuffer("");
    while ((len = InStr.read(buf)) != 2) {
        System.out.println(len);
        StrBuf.append(new String(buf, 0, len));
    }
    InStr.close();
    System.out.println(StrBuf.toString());
}
```

控制台的输出由 print 和 println()方法完成，这些方法都由 PrintStream 类定义，System.out 是 PrintStream 类对象的一个引用。PrintStream 继承自 OutputStream 类，并且实现了 write 方法，write 方法也可以用来在控制台进行写操作。

2．与文件相关的类

与文件相关的类如下：

（1）File 类：以抽象的方式表示文件名或目录名，该类主要用于文件和目录的创建、文件的查找和删除等。

（2）FileReader 类：从 InputStreamReader 类继承而来，该类按字符读取流中数据。

（3）FileWriter 类：从 OutputStreamWriter 类继承而来，该类按字符向流中写入数据。

File 类的部分方法如表 4.6 所示。

表 4.6　File 类的部分方法

方　法	说　明
public String getName()	返回由抽象目录名表示的文件或目录的名称
public String getParent()	返回此抽象目录名的父目录名的目录名字符串，如果此目录名没有指定父目录，则返回 null
public File getParentFile()	返回此抽象目录名的父目录名的抽象目录名，如果此目录名没有指定父目录，则返回 null
public String getPath()	将此抽象目录名转换为一个目录名字符串
public boolean isAbsolute()	测试此抽象目录名是否是绝对目录名
public String getAbsolutePath()	返回抽象目录名的绝对目录名字符串

续表

方　法	说　明
public boolean canRead()	测试应用程序是否可以读取此抽象目录名表示的文件
public boolean canWrite()	测试应用程序是否可以修改此抽象目录名表示的文件
public boolean exists()	测试此抽象目录名表示的文件或目录是否存在
public boolean isDirectory()	测试此抽象目录名表示的文件是否是一个目录
public boolean isFile()	测试此抽象目录名表示的文件是否是一个标准文件

4.4　开发实践：温度和湿度数据的记录

4.4.1　开发设计

本任务通过一个 Java 程序来计算最近一周温度和湿度的平均值。在程序中定义了 DataGenerater 类，用于产生最近一周的温度和湿度数据。程序中模拟的数据如图 4.2 所示，程序执行流程如图 4.3 所示。

```
package com.example.test.libtest;

public class DataGenerater {
    /** 返回最近一周的温度*/
    public static String getTemperatures() {
        return "21, 22, 19, 28, 29, 32, 35";
    }

    /*** 返回最近一周的湿度*/
    public static String getHumiditys() {
        return "0.45, 0.23, 0.58, 0.32, 0.44, 0.56, 0.48";
    }
}
```

图 4.2　程序中模拟的数据

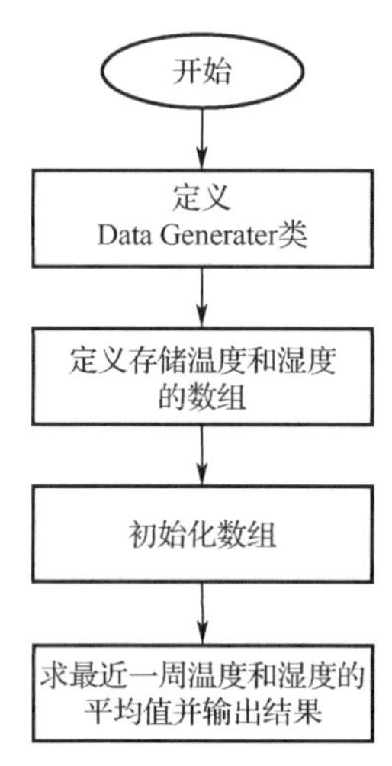

图 4.3　程序执行流程

4.4.2　功能实现

本任务的目标是计算最近一周温度和湿度的平均值，首先定义一个数组来保存最近一周的温度和湿度，然后取出数组中的数据进行累加求平均值。

1．产生最近一周的温度和湿度

```
public class DataGenerater {
    /*返回最近一周的温度*/
    public static String getTemperatures() {
        return "21, 22, 19, 28, 29, 32, 35";
```

```
    }
    /*返回最近一周的湿度*/
    public static String getHumiditys() {
        return "0.45, 0.23, 0.58, 0.32, 0.44, 0.56, 0.48";
    }
}
```

2．定义并初始化存储温度和湿度的数组

```
int temperatureArray[] = new int[7];
float humidityArray[] = new float[7];
```

通过 DataGenerater 类的 getTemperatures 方法与 getHumiditys 方法来初始化存储温度和湿度的数组。

```
//记录温度
String[] s1 = DataGenerater.getTemperatures().split(",");
for (int i = 0; i < 7; i++) {
    temperatureArray[i] = Integer.valueOf(s1[i].trim());
}
//记录湿度
String[] s2 = DataGenerater.getHumiditys().split(",");
for (int i = 0; i < 7; i++) {
    humidityArray[i] = Float.valueOf(s2[i].trim());
}
```

3．求最近一周温度和湿度的平均值并输出结果

```
//求最近一周温度的平均值并输出结果
int sum1 = 0;
for(int i = 0; i < 7; i++) {
    sum1 = sum1 + temperatureArray[i];
}
System.out.println("最近一周温度平均值："+sum1/7);
//求最近一周湿度的平均值并输出结果
float sum2 = 0;
for(int i = 0; i < 7; i++) {
    sum2 = sum2 + humidityArray[i];
}
System.out.println("最近一周湿度平均值："+sum2/7);
```

4.5　任务验证

首先将开发例程导入 Android Studio 开发环境中并定位到 TestClass 文件，然后运行该文件，如图 4.4 所示。

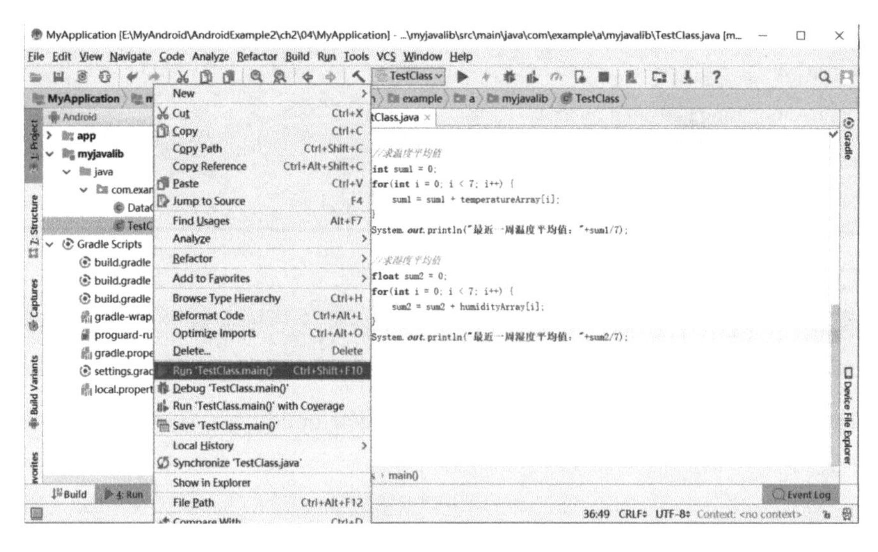

图 4.4　运行 TestClass 文件

控制台会输出最近一周温度和湿度的平均值，如图 4.5 所示。

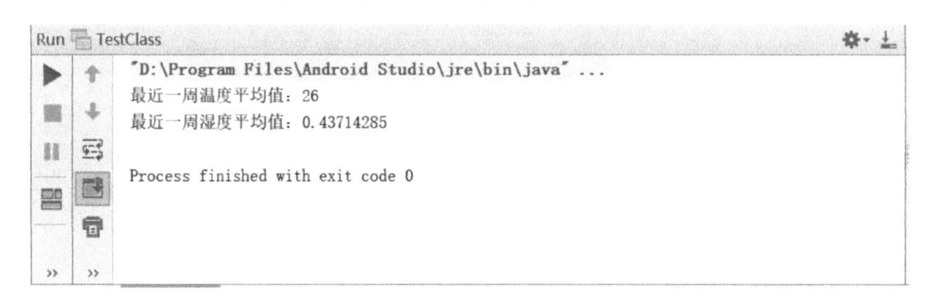

图 4.5　程序输出结果

4.6　开发小结

本任务重点介绍了 Java 的常用类、Java 数组、Java 方法、Java IO 流，可以帮助读者掌握 Java 常用类、数组、方法与 IO 流的概念与基本使用。本任务最后给出了一个开发实践，读者通过该开发实践可以初步掌握 Java 编程的技能，实现温度和湿度平均值的计算。

4.7　思考与拓展

（1）Java 中的 int 和 integer 有什么区别？

（2）请尝试修改本任务的例程，找到最近一周温度和湿度的最大值。

任务 5

智能背包系统的设计

本任务介绍 Java 的类和对象，以及类的继承与封装，深入讲解 Java 的接口实现和包机制等面向对象编程的特征。

5.1 开发场景：如何使用 Java 封装一个智能背包系统

户外登山爱好者的设备越来越智能化，如智能背包，可以随时记录海拔高度、温度、湿度等。本任务使用 Java 将智能背包系统封装成一个类。

5.2 开发目标

（1）知识目标：熟悉 Java 对象与类的定义、Java 的常用类、类的继承和封装、Java 的接口。

（2）技能目标：熟悉 Java 类的继承与封装、Java 的接口实现和包机制。

（3）任务目标：掌握 Java 类的创建方法，能够动手封装实际生活中的类，将智能背包系统封装成一个类。

5.3 原理学习：Java 的对象与类、接口实现与包机制

5.3.1 Java 的对象与类

1. 对象与类的定义

（1）Java 的对象。现实世界有很多对象，如猫、狗等，这些对象都有自己的状态和行为，如猫的状态有名字、品种和颜色等，猫的行为有叫、摇尾巴和跑等。

软件对象也有状态和行为。软件对象的状态就是属性，行为可以通过方法来实现。在软件开发中，方法可以改变对象内部的状态，对象之间的相互调用也是通过方法来完成的。

（2）Java 的类。类可以看成创建对象的模板，一个类通常包含以下类型变量：

① 局部变量：在方法、构造函数或者语句块中定义的变量被称为局部变量，变量声明和初始化都是在方法中实现的，方法结束后，局部变量会被自动销毁。

② 成员变量：成员变量是定义在类中、方法外的变量。成员变量在创建对象时被实例化，成员变量可以被类中方法、构造函数和特定类的语句块访问。

③ 类变量：类变量在类中、方法外声明，必须声明为静态类型。

（3）定义构造函数。每个类都有构造函数。如果没有显式地为类定义构造函数，则Java编译器会为该类提供一个默认构造函数。在创建一个对象时，至少要调用一个构造函数。构造函数的名称必须与类名相同，一个类可以有多个构造函数。下面是一个构造函数的示例。

```
public class Dog{
    public Dog(){
    }
    public Dog(String name){
        //这个构造函数仅有一个参数，即 name
    }
}
```

（4）创建对象。对象是根据类创建的，在Java中，可以使用关键字new来创建一个对象。创建对象需要以下三步：

① 声明：声明一个对象，包括对象名称和对象类型。

② 实例化：使用new创建一个对象。

③ 初始化：在使用new创建对象时，会调用构造函数来初始化对象。

下面是一个创建对象的示例。

```
public class Dog {
   public Dog(String name){
      //这个构造函数仅有一个参数，即 name
      System.out.println("小狗的名字是 : " + name );
   }
   public static void main(String []args){
      //下面的语句将创建一个 Dog 对象
      Dog myDog = new Dog( "Rafael" );
   }
}
```

（5）访问变量和方法。在Java中，可以通过已创建的对象来访问类中的变量和方法，如下所示。

```
/*实例化对象*/
Object objectName= new Object();
/*访问类中的变量*/
objectName.variableName;
/*访问类中的方法*/
objectName.methodName();
```

（6）源文件的声明规则。如果要在一个源文件中定义多个类，并且还有 import 语句和 package 语句，则应遵守以下声明规则：

① 一个源文件中只能有一个 public 类。

② 一个源文件可以有多个非 public 类。

③ 源文件的名称应该和 public 类的名称保持一致。例如，若源文件中 public 类的名称是 Dog，则源文件应该命名为 Dog.java。

④ 如果一个类定义在某个包中，则 package 语句应该在源文件的首行。

⑤ 如果源文件包含 import 语句，则该语句应该放在 package 语句和类定义之间；如果没有 package 语句，则 import 语句应该放在源文件中最前面。

⑥ import 语句和 package 语句在源文件中定义的所有类都有效。

（7）Java 包。将功能相似或相关的类和接口放在同一个包中，可以方便类和接口的查找及使用，对类和接口进行分类。当开发 Java 程序时，可能会涉及很多类和接口，因此有必要对类和接口进行分类。

（8）import 语句。import 语句用来提供一个合理的目录，使得编译器可以找到某个类。例如，采用下面的语句可以载入 Java 安装目录“/java/io”下的所有类。

```
import java.io.*;
```

2. Java 常用类介绍

（1）Java 类库的结构。类库就是 Java API，是系统提供的标准类的集合。Java 类库中的类和接口大多封装在特定的包里，每个包具有自己的功能。表 5.1 列出了 Java 中的常用包，其中，包名后面带“.*”的表示其中包括一些相关的包，通过查阅 Java 技术文档可以了解更多的类库。

表 5.1 Java 中的常用包

包 名	主要功能	包 名	主要功能
java.applet	提供了创建 Applet 所需要的类	java.awt.*	提供了创建用户界面，以及绘制和管理图形、图像的类
java.beans.*	提供了开发 Java Beans 所需要的类	java.io	提供了通过数据流、对象序列，以及文件系统实现的系统输入、输出的类
java.lang.*	Java 的基本类库	java.math.*	提供了简单的整数算术及十进制算术的基本函数类
java.rmi	提供了与远程方法调用相关的类	java.net	提供了用于实现网络通信应用的类
java.security.*	提供了设计网络安全方案所需要的类	java.sql	提供了访问和处理来自 Java 标准数据源数据的类
java.test	包括以一种独立于自然语言的方式处理文本、日期、数字和消息的类与接口	java.util.*	包括集合类、时间处理模式、日期时间工具等各类常用工具包
javax.accessibility	定义了在用户界面组件之间进行相互访问的一种机制	javax.naming.*	为命名服务提供了一系列类和接口
javax.swing.*	提供了一系列轻量级的用户界面组件，是目前 Java 用户界面常用的包之一	javax.swing.*	提供了一系列轻量级的用户界面组件，是目前 Java 用户界面常用的包之一

除了 java.lang，其他的包都需要用 import 语句导入之后才能使用。

（2）基本数据类型的类。java.lang 不仅包含了 Object 类，java.lang.Object 类是 Java 中整

个类层次结构的根节点，还定义了基本数据类型的类，如 Boolean、Character、Byte、Integer、Short、Long、Float 和 Double 等，这些类支持基本数据类型的转换，如表 5.2 所示。

表 5.2　基本数据类型的类

基本数据类型	对应的包装类	基本数据类型	对应的包装类
byte	Byte	short	Short
int	Integer	long	Long
float	Float	double	Double
char	Character	boolean	Boolean

（3）String 类和 StringBuffer 类。Java 字符串是由 String 类和 StringBuffer 类来处理的。Java 中的字符串属于 String 类，虽然可以用其他方法表示字符串，但使用 String 类作为字符串的标准格式，Java 编译器可以把字符串转换成 String 对象。如果需要进行大量的字符串操作，就可以使用 StringBuffer 类或字符数组。

StringBuffer 类与 String 类相似，它具有 String 类的很多功能。StringBuffer 对象可以在缓冲区内被修改，如增加、替换字符或子字符串。在完成缓冲字符串数据操作后，可以通过 StringBuffer.toString 方法或 String 类的构造函数将字符串转换成标准字符串格式。

（4）System 类。System 类是一个特殊类，它是一个公共最终类，不能被继承，也不能被实例化，即不能创建 System 对象。System 类功能强大，与 Java 运行时一起可以访问许多有用的系统功能。System 类是保存静态方法和变量的集合，标准的输入、输出和 Java 运行时的错误输出都存储在变量 in、out 和 err 中。System 类中所有的变量和方法都是静态的，使用时以 System 作为前缀，如“System.变量名”和“System.方法名”。

（5）Math 类。Math 类提供了用于基本数学运算的属性和方法。

（6）Vector 类。当创建过大的数组时，会造成空间的浪费。Java 中的 java.util 包提供了 Vector 类，通过该类可以根据需要创建动态数组。另外，Vector 类还提供了一些有用的方法，如增加和删除元素的方法。

（7）Stack 类。Stack 类是 Vector 类的一个子类，用于实现后进先出的堆栈。Stack 类定义了创建空堆栈的构造函数，不仅包括由 Vector 类定义的所有方法，还增加了几种它自己定义的方法。

（8）ArrayList 类。ArrayList 是一个容量能够动态变化的动态数组，它继承自 AbstractList 类，实现了 List、RandomAccess、Cloneable、java.io.Serializable 等接口。

ArrayList 类适合随机访问元素，但在数组中插入和删除元素的效率较低。由于 ArrayList 类不是线程安全的，因此该类适合在单线程中使用，而在多线程中则使用 Vector 类或者 CopyOnWriteArrayList 类。

（9）HashMap 类和 HashSet 类。HashMap 类和 HashSet 类是 Java 集合框架（Collection Framework）中的两个重要成员，HashMap 是 Map 接口的常用实现类，HashSet 是 Set 接口的常用实现类。

3. 继承

继承允许创建不同等级的类，子类可以继承父类的特征和行为，从而使得子类的对象

（实例）具有父类的实例域和方法。子类也可以从父类继承方法，使得子类与父类具有相同的行为。

Java 可以通过关键字 extends 来声明一个类是从另外一个类继承而来的，格式如下：

```
class 父类 {
}
class 子类 extends 父类 {
}
```

（1）继承的特性。

① 子类拥有父类非 private 的属性和方法。

② 子类可以拥有自己的属性和方法，即子类可以对父类进行扩展。

③ 子类可以用自己的方式实现父类的方法。

④ Java 的继承是单继承，但是可以多重继承。单继承是指一个子类只能继承一个父类；多重继承是指多层次的继承，如 B 类继承 A 类、C 类继承 B 类，按照关系来说，B 类是 C 类的父类，A 类是 B 类的父类。

⑤ 继承可以提高类之间的耦合性，但耦合度高就会造成代码之间的联系变得紧密，使独立性变差。

（2）继承的关键字。Java 可以通过关键字 extends 和 implements 来实现继承。在 Java 中，所有的类都继承自 Object，如果一个类没有使用继承的两个关键字，则默认为继承自 Object 类。

① extends 关键字。类的继承是单一继承，一个子类只能拥有一个父类，所以 extends 只能继承自一个类。

② implements 关键字。使用 implements 关键字可以变相地使 Java 具有多继承的特性，适用于继承接口的情况，可以同时继承多个接口。

③ super 与 this 关键字。通过 super 关键字可以访问父类成员，用来引用当前对象的父类；this 关键字用于指向自己的引用。

④ final 关键字。使用 final 关键字声明的类是不能被继承的，为最终类；使用 final 关键字修饰的类的方法不能被该类的子类覆写；定义为 final 的实例变量不能被修改；被声明为 final 类，其方法将会被自动声明为 final，但实例变量并不是 final。

（3）构造函数。子类不能继承父类的构造函数。如果父类的构造函数带有参数，则必须在子类的构造函数中显式地通过 super 关键字调用父类的构造函数并配以适当的参数列表；如果父类的构造函数没有参数，则在子类的构造函数中通过 super 关键字调用父类构造函数不是必需的，如果没有使用 super 关键字，系统会自动调用父类的无参构造函数。

4．封装

封装是一种将抽象性函数接口的实现细节部分包装、隐藏起来的方法。适当的封装可以让代码变得更容易理解与维护。

（1）封装的优点。

① 良好的封装能够减少耦合。

② 可以对成员变量进行更精确的控制。

③ 可以隐藏实现细节。

（2）实现 Java 封装的步骤

① 通过修改属性的可见性来限制对属性的访问（一般限制为 private）。例如：

```
public class Student {
    private String name;
    private int age;
}
```

② 对每个属性提供对外的公共方法，也就是创建一对赋/取值方法，用于对私有属性进行访问。例如：

```
public class Student{
    private String name;
    private int age;
    public int getAge(){
        return age;
    }
    public String getName(){
        return name;
    }
    public void setAge(int age){
        this.age = age;
    }
    public void setName(String name){
        this.name = name;
    }
}
```

5.3.2 Java 的接口

1．接口的定义

接口是一个抽象类型，是抽象方法的集合，接口通常用 interface 关键字来声明。一个类可以通过继承接口的方式来继承接口的抽象方法。

接口并不是类，编写接口的方式和类很相似，但是它们属于不同的概念。类描述的是对象的属性和方法，接口则包含类要实现的方法。除非实现接口的类是抽象类，否则该类要定义接口中的所有方法。接口无法被实例化，但是可以被实现。在类实现接口时，必须实现接口内所描述的所有方法，否则就必须声明为抽象类。另外，在 Java 中，接口类型可用来声明一个变量。

2．接口的声明

interface 关键字用来声明一个接口，格式如下：

```
[可见度] interface 接口名称 [extends 其他的类名] {
    //声明变量
    //抽象方法
}
```

接口具有以下特性：

（1）接口是隐式抽象的，当声明一个接口时，不必使用 abstract 关键字。

（2）接口中每一个方法也是隐式抽象的，在声明接口中的方法时同样也不必使用 abstract 关键字。

（3）接口中的方法都是公有的。

3. 接口的实现

当类实现接口时，类要实现接口中所有的方法，否则该类必须声明为抽象类。类使用 implements 关键字实现接口，在类声明中，implements 关键字应放在 class 声明后面。实现一个接口的格式如下：

```
…implements 接口名称[其他接口名称, 其他接口名称…, …] ….
```

例如：

```
public class Dog implements Animal{
    public void eat(){
        System.out.println("Dog eats");
    }
    public void travel(){
        System.out.println("Dog travels");
    }
    public int noOfLegs(){
        return 4;
    }
    public static void main(String args[]){
        Dog dog = new Dog();
        dog.eat();
        dog.travel();
    }
}
//结果输出
Dog eats
Dog travels
```

在覆写接口中声明的方法时，需要注意以下规则：

① 类在实现接口的方法时，不能抛出强制性异常，只能在接口中或者继承接口的抽象类中抛出该强制性异常。

② 类在覆写方法时要保持方法名的一致，并且应该保持相同或者相兼容的返回值类型。

③ 如果实现接口的类是抽象类，就没必要实现该接口的方法。

在实现接口时，也要注意以下规则：

① 一个类可以同时实现多个接口。

② 一个类只能继承一个类，但能实现多个接口。

③ 一个接口能继承另一个接口。

4．接口的继承

一个接口能继承另一个接口，接口的继承使用 extends 关键字。下面的 Animal 接口被 Bird 和 Fish 接口继承：

```
//文件名：animal.java
public interface Animal
{
    public void eat( );
    public void sleep();
}
//文件名：Bird.java
public interface Bird extends Animal
{
    public void fly();
}
//文件名:Fish.java
public interface Fish extends Animal
{
    public void swim ();
}
```

Bird 接口声明了 1 个方法，从 Animal 接口继承了 2 个方法，这样在实现 Bird 接口的类时需要实现 3 个方法。

5．接口的多继承

类的多继承是不合法的，但接口允许多继承。在接口的多继承中，extends 关键字只需要使用一次，在其后跟着继承接口即可。例如：

```
public interface Huckword extends Swim, Play
```

上述语句定义的子接口是合法的。与类不同的是，接口允许多继承。

6．标记接口

最常用的继承接口是不包含任何方法的接口，不包含任何方法的接口称为标记接口。标记接口是一种不包含任何方法和属性的接口，仅仅表明它的类属于一个特定的类型。标记接口的作用是给某个对象打个“标”，使对象拥有某个或某些特权。例如，java.awt.event 包中的 MouseListener 接口继承自 java.util.EventListener 接口，定义如下：

```
package java.util;
public interface EventListener
{
}
```

标记接口主要用于以下两种场合。

（1）建立一个公共的父接口：如 EventListener 接口，这是由几十个其他接口扩展的 Java API，可以使用一个标记接口来建立一组接口的父接口。

（2）向一个类添加数据类型：实现标记接口的类不需要定义任何接口方法，但该类可通过多态性变成一个接口类型。

5.4　开发实践：智能背包设备系统对象的设计

5.4.1　开发设计

本任务利用 Java 面向对象的特点，将智能背包系统封装为 Java 的一个类，即 SmartBag 类。智能背包系统的管理功能（如智能背包系统设备的开启和关闭，以及状态的报告等）全部封装在类的包中。SmartBag 类的定义如图 5.1 所示。

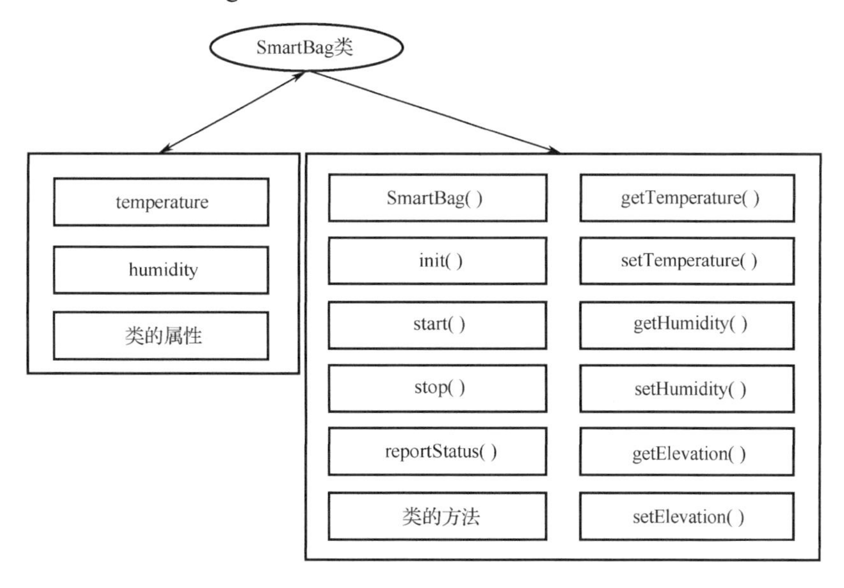

图 5.1　SmartBag 类的定义

5.4.2　功能实现

根据开发设计，首先定义 SmartBag 类，然后实现 SmartBag 类的功能，最后在 main 函数中新建 SmartBag 类的对象，并验证功能。代码如下：

1. 定义 SmartBag 类并实现其功能

```
package com.example.a.myjavalib;
public class SmartBag {
    //温度
    private float temperature;
    //湿度
    private float humidity;
    //海拔
    private int elevation;
```

```
    public SmartBag() {
        System.out.println("Hi,我是一个智能背包设备。");
    }
    public float getTemperature() {
        return temperature;
    }
    public void setTemperature(float temperature) {
        this.temperature = temperature;
    }
    public float getHumidity() {
        return humidity;
    }
    public void setHumidity(float humidity) {
        this.humidity = humidity;
    }
    public int getElevation() {
        return elevation;
    }
    public void setElevation(int elevation) {
        this.elevation = elevation;
    }
    private void init() {
        System.out.println("初始化设备。");
        temperature = 26;
        humidity = 0.45f;
        elevation = 100;
    }
    public void start() {
        init();
        System.out.println("设备成功开启。");
    }
    public void stop() {
        System.out.println("设备成功关闭。");
    }
    public void reportStatus() {
        String status =  "当前温度温："+temperature+"，当前湿度："+humidity+"，当前海拔："+elevation;
        System.out.println(status);
    }
}
```

2．在 main 函数中新建 SmartBag 类的对象并验证功能

```
public class TestClass {
    public static void main(String[] args) {
        //新建一个 SmartBag 类的对象
        SmartBag device = new SmartBag();
        //启动设备
```

```
            device.start();
            //报告状态
            device.reportStatus();
            //关闭设备
            device.stop();
    }
}
```

5.5 任务验证

在 Android Studio 开发环境打开本任务的例程并定位到 TestClass 文件，然后运行该文件，如图 5.2 所示。

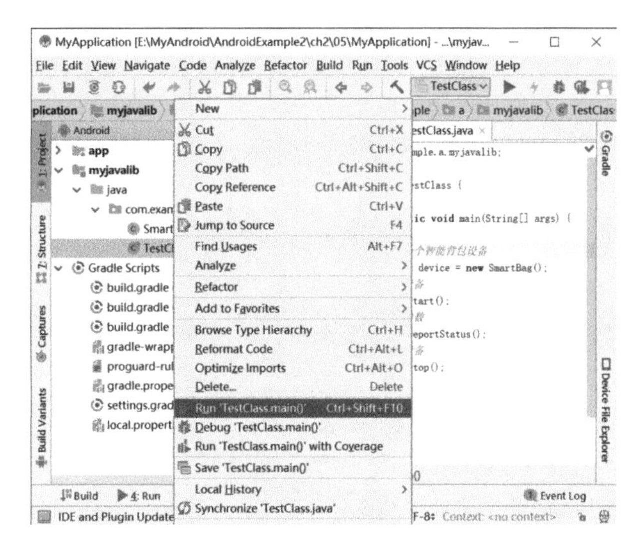

图 5.2 运行文件

控制台会输出程序运行结果，如图 5.3 所示。

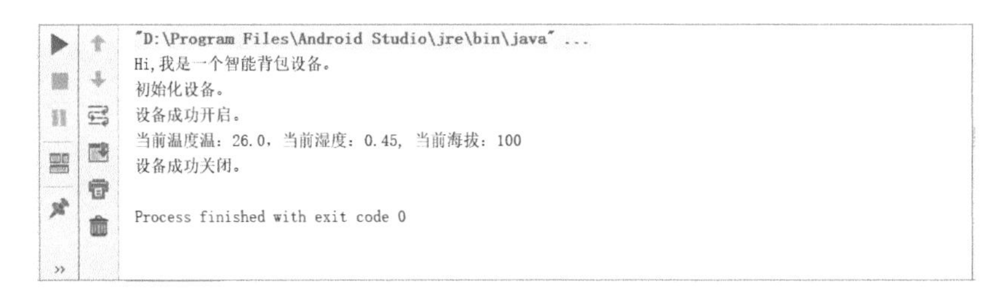

图 5.3 程序运行结果

5.6 开发小结

本任务通过介绍 Java 的类和对象来初步阐述面向对象编程的思想；通过介绍 Java 类的

封装与继承，以及 Java 的接口实现来进一步加深读者对 Java 特性的理解；通过创建智能背包系统对象来加深读者对 Java 面向对象编程思想的理解。

5.7 思考与拓展

（1）在 Java 中，一个类最多可以继承多少个父类？最多可以实现多少个接口？

（2）请尝试同时创建两个智能背包系统对象，并输出状态。

第 2 篇

Android 开发基础

本篇主要介绍 Android 的基本知识和语法，通过开发实践帮助读者熟悉 Android 的开发基础。本篇共 5 个任务：

任务 6 为 Android 项目框架与调试。

任务 7 为空气质量显示界面的设计。

任务 8 为城市气象监控设备管理系统应用界面的设计。

任务 9 为城市环境系统功能菜单的设计。

任务 10 为城市灯光控制系统界面事件的处理。

任务 6 Android 项目框架与调试

本任务主要介绍 Android 项目框架，以及项目的调试。通过本任务的学习，读者可以掌握调试工具的使用，以及模拟器的安装与使用，并创建和调试 Android 项目。

6.1 开发场景：如何创建和调试 Android 项目

任务 1 介绍了 Android Studio 开发环境的搭建，本书中的实例均是在 Android Studio 开发环境中进行开发的。Android Studio 开发环境界面如图 6.1 所示。

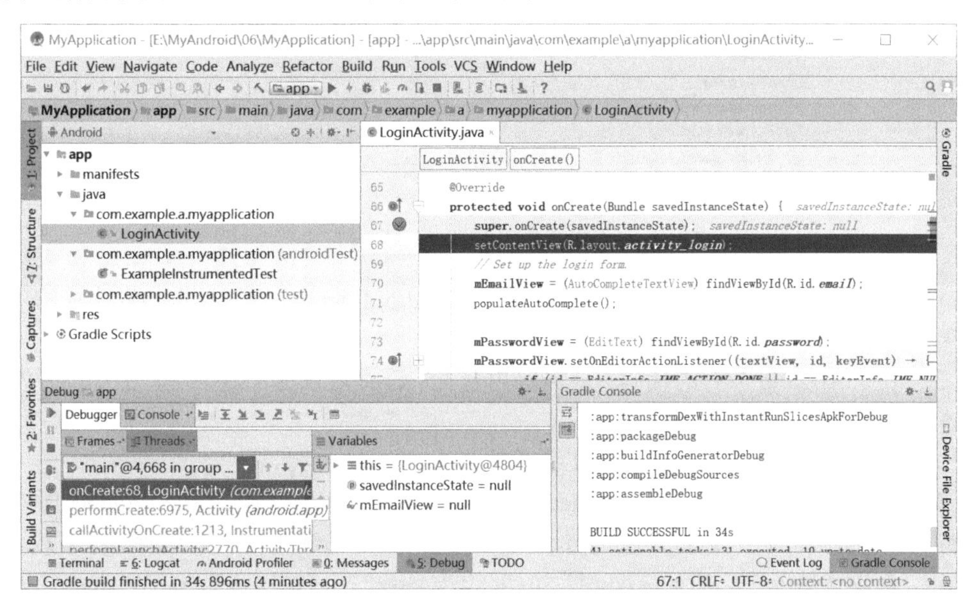

图 6.1 Android Studio 开发环境界面

6.2 开发目标

（1）知识目标：熟悉 Android 项目框架，掌握模拟器的安装与使用，以及调试工具的使用。

（2）技能目标：掌握 Android 项目的创建和调试。

（3）任务目标：熟悉 Android 项目目录结构及项目运行，掌握模拟器的安装与使用，以及调试工具的使用，能够创建和调试 Android 项目。

6.3 原理学习：Android 项目框架及项目创建调试

6.3.1 Android 项目框架

1. Android 项目目录结构

理解 Android 项目目录结构很重要，读者需要很清楚每个目录文件有什么作用、什么时候用、有哪些资源、放在什么地方、如何增加删除更新。一个比较简单的 Android 项目目录结构如图 6.2 所示。

图 6.2　一个比较简单的 Android 项目目录结构

（1）.gradle 和.idea。这两个目录下存放的是 Android Studio 开发环境自动生成的一些文件，无须关心，也不要去手动修改。

（2）app。Android 项目的代码、资源等内容几乎都存放在这个目录下，后面的开发工作也基本都是在这个目录下进行的。

（3）build。该目录主要存放一些在编译时自动生成的文件。

（4）gradle。该目录用于存放 Gradle Wrapper 的配置文件，使用 Gradle Wrapper 时不需要提前将 Gradle Wrapper 下载好，系统会自动根据本地的缓存情况决定是否需要连网下载 Gradle Wrapper。在默认情况下，启动 Android Studio 开发环境时不会打开 Gradle Wrapper，如果需要，则可以通过单击 Android Studio 开发环境的菜单“File→Settings→Build, Execution, Deployment→Gradle”来打开 Gradle Wrapper。

① .gitignore。该文件用来将指定的目录或文件排除在版本控制之外。

② build.gradle。该文件是项目全局的 gradle 构建脚本，通常这个文件中的内容是不需要修改的。

③ gradle.properties。该文件是全局的 gradle 配置文件，在这里配置的属性将会影响到项

目中所有的 Gradle Wrapper 编译脚本。

④ gradlew 和 gradlew.bat。这两个文件用来在命令行窗口中执行 Gradle Wrapper 命令，其中 gradlew 文件在 Linux 或 Mac 系统中使用的，gradlew.bat 文件在 Windows 系统中使用的。

⑤ local.properties。该文件用于指定本机中的 Android SDK 路径，内容都是自动生成的，不需要修改。当 Android SDK 位置发生了变化，将这个文件中的路径改成新的路径即可。

⑥ MyApplication.iml。该文件是所有 IntelliJ IDEA 项目都会自动生成的一个文件（Android Studio 是基于 IntelliJ IDEA 开发的），用于标识这是一个 IntelliJ IDEA 项目，不需要修改这个文件中的内容。

⑦ settings.gradle。该文件用于指定项目中所有引入的模块。由于 MyApplication 项目中只有一个 app 模块，因此该文件中也就只引入了 app 这一个模块，模块的引入是自动完成的。

2．Android 项目运行解析

（1）Android 项目的启动方式。Android 项目的启动方式有以下两种：

① 冷启动：当启动 Android 项目时，如果后台没有该项目的进程，则系统会新建一个进程分配给该项目。在冷启动时，系统会新建一个进程，首先会创建和初始化 Application 类，然后创建和初始化 MainActivity 类，最后显示在界面上。

② 热启动：当启动 Android 项目时，如果后台已有该项目的进程，则会从已有的进程中来启动该项目。热启动是从已有的进程中来启动 Android 项目的，只需要创建和初始化一个 MainActivity 类即可，不必创建和初始化 Application 类。一个 Android 项目从创建进程到销毁进程，Application 类只需要初始化一次。

（2）Android 项目的特点。

① 每个 Android 项目都在一个独立空间里运行，这意味着其运行在一个单独的进程中，拥有自己的虚拟机，系统为 Android 项目分配了一个唯一的用户 ID。

② Android 项目由多个组件组成，这些组件还可以启动其他项目的组件，因此，Android 项目并没有一个类似程序入口的 main()函数。

（3）Android 项目的运行。Android 进程与 Linux 进程一样，每个 apk 运行在自己的进程中。另外，在默认情况下，一个进程里面只有一个线程，即主线程，这个主线程中有一个 Looper 实例，通过调用 Looper.loop()函数可从消息队列里面取出消息来做相应的处理。Android 项目运行流程 6.3 所示。

Android 项目的运行流程可以分为以下三步：

① 创建进程。用户单击 Android 项目图标时，Click 事件会调用 startActivity(intent)，通过 Blinder IPC 机制来调用 ActivityManagerService，通过 startProcessLocked()方法来创建新的进程，该方法会通过 Socket 将传递参数给 Zygote 进程。Zygote 进程调用 ZygoteInit.main()方法来实例化 ActivityThread 对象并最终返回新进程的 PID。ActivityThread 对象依次调用 Looper.prepareLoop()和 Looper.loop()来开启消息循环。

② 绑定 Android 项目。首先通过 ActivityThread 对象调用 bindApplication()方法可以将进程和指定的 Application 绑定起来，该方法会发送一个 BIND_APPLICATION 消息到消息队列中；然后最终通过 handleBindApplication()方法处理 BIND_APPLICATION 消息；最后调用 makeApplication()方法来将 Android 项目中的类加载到内存中。

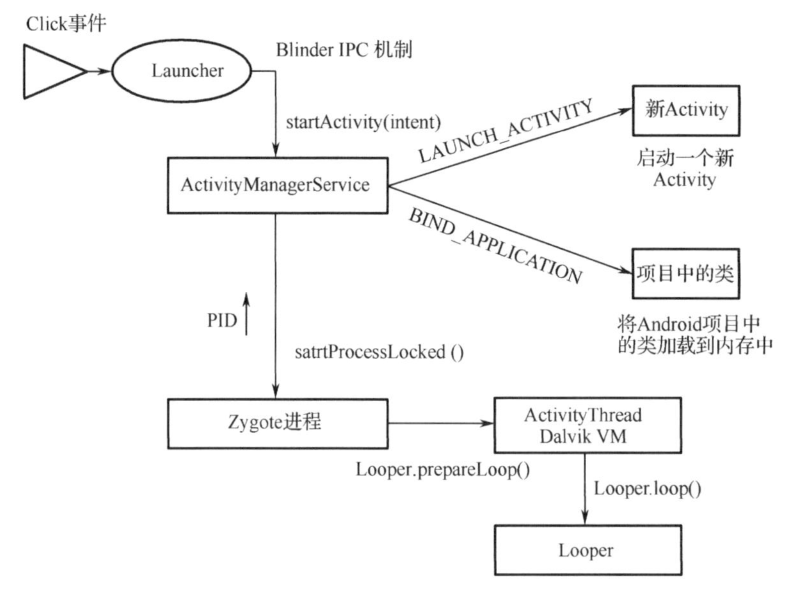

图 6.3　Android 项目运行流程

③ 启动新 Activity。此时系统已经拥有了 Android 项目的进程，这时就可以从一个已经存在的进程中启动一个新 Activity。

6.3.2　Android 项目的调试

调试可以快速地查找 Android 项目出现的问题，本节简要地介绍调试的基本使用方法和技巧。

1．进入调试

（1）设置断点。选定要设置断点的代码行，在行号后面单击鼠标左键即可设置断点，如图 6.4 所示。

```
@Override
protected void onCreate(Bundle savedInstanceState) {
    super.onCreate(savedInstanceState);
    setContentView(R.layout.activity_login);
    // Set up the login form.
    mEmailView = (AutoCompleteTextView) findViewById(R.id.email);
    populateAutoComplete();

    mPasswordView = (EditText) findViewById(R.id.password);
    mPasswordView.setOnEditorActionListener(new TextView.OnEditorActionListener() {
        @Override
        public boolean onEditorAction(TextView textView, int id, KeyEvent keyEvent) {
            if (id == EditorInfo.IME_ACTION_DONE || id == EditorInfo.IME_NULL) {
                attemptLogin();
                return true;
            }
            return false;
        }
    });
```

图 6.4　设置断点

（2）进入调试状态。在设置断点后，单击工具栏中的 Debug 按钮即可进入调试状态，如图 6.5 所示。

图 6.5　进入调试状态

当 Android 项目进入调试状态后，Android Studio 开发环境会弹出 Debug 窗口，如图 6.6 所示，即调试者状态，可以对的程序进行监视和调试。

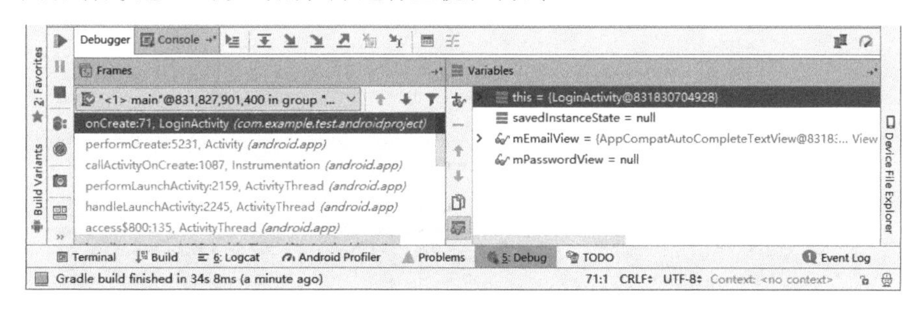

图 6.6　Debug 窗口

2. 常用的调试功能

（1）step into：单步执行，在遇到子函数时会进入子函数并继续单步执行。快捷键为 F7。

（2）step over：单步执行，在遇到子函数时会进入子函数内，但不再单步执行，而是将子函数整个执行完再停止，也就是把整个子函数作为一步。快捷键为 F8。

（3）step out：在子函数内单步执行时，使用该功能可以执行完子函数的剩余部分，并返回到上一层函数。快捷键为 Shift+F8。

（4）Run to Cursor：执行该功能后，不论程序执行到哪里，都可以执行到光标所在的行。快捷键为 Alt+F9。

（5）show Execution Point：当不知道程序执行到哪里时，通过该功能可以使 Android Studio 开发环境跳到执行行所在的界面，并将该行高亮显示出来。快捷键为 Alt+F10。

6.3.3　模拟器的安装与使用

如果在运行 Android 项目时弹出如图 6.7 所示的“Select Deployment Target”对话框，这表示没有连接可用的设备，这时就需要创建一个模拟器。勾选图 6.7 中的“Use same selection for future launches”，然后单击“OK”按钮即可创建一个模拟器。

创建模拟器时需要设置模拟硬件的型号，如图 6.8 所示。

设置模拟硬件的型号后还需要选择 Android 系统镜像，如图 6.9 所示。

接下来还需要配置 Android 虚拟设备，如图 6.10 所示。

Android 虚拟设备配置完成后的运行效果如图 6.11 所示。在运行 Android 项目时，就会在模拟器上显示相应的界面。

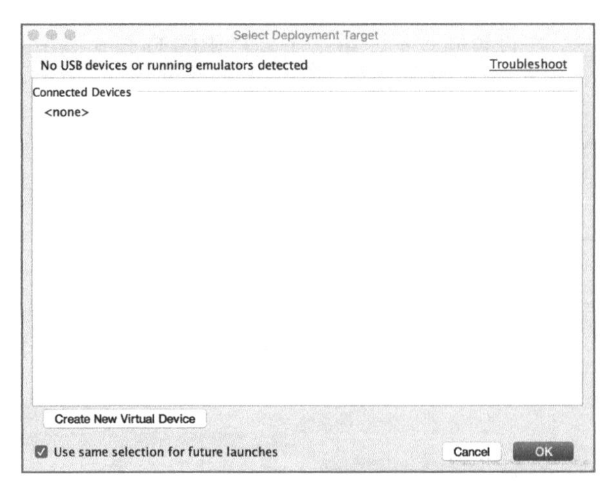

图 6.7 “Select Deployment Target”对话框

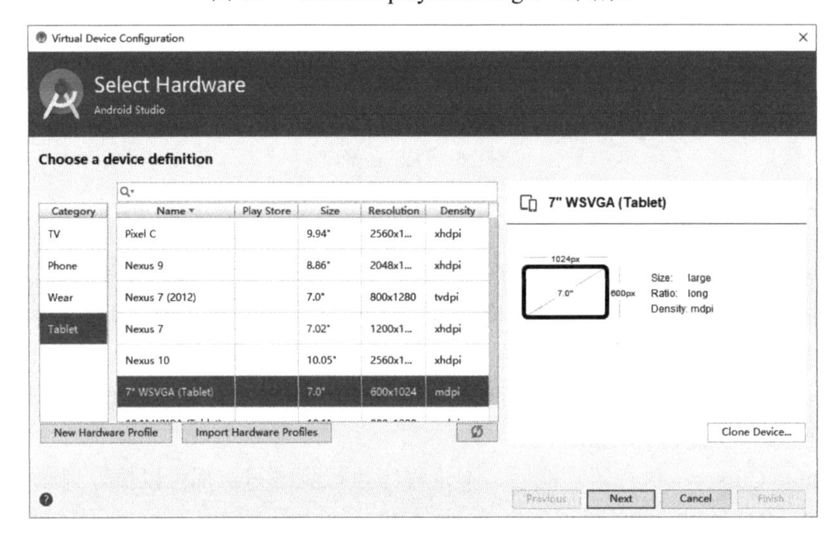

图 6.8 设置模拟硬件的型号

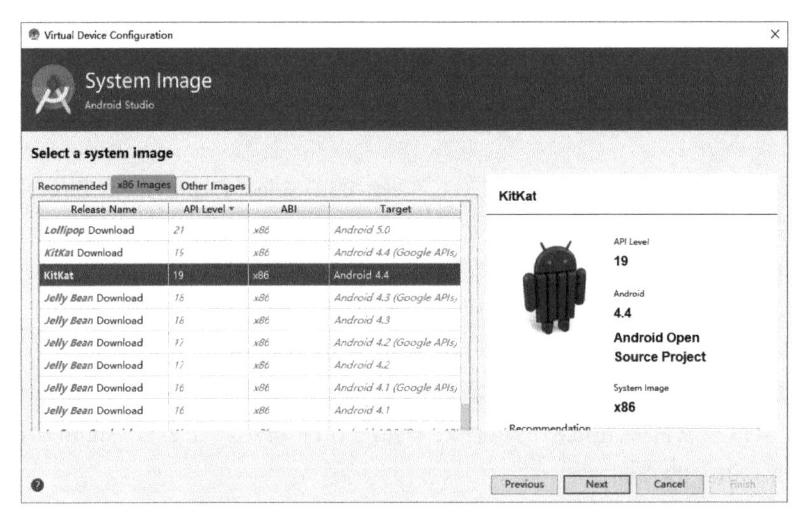

图 6.9 选择 Android 系统镜像

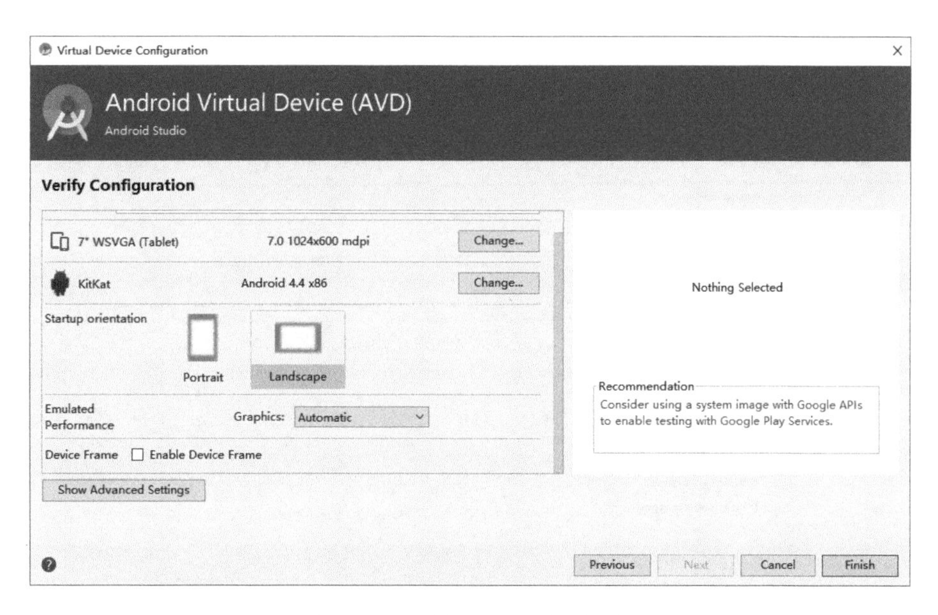

图 6.10　配置 Android 虚拟设备

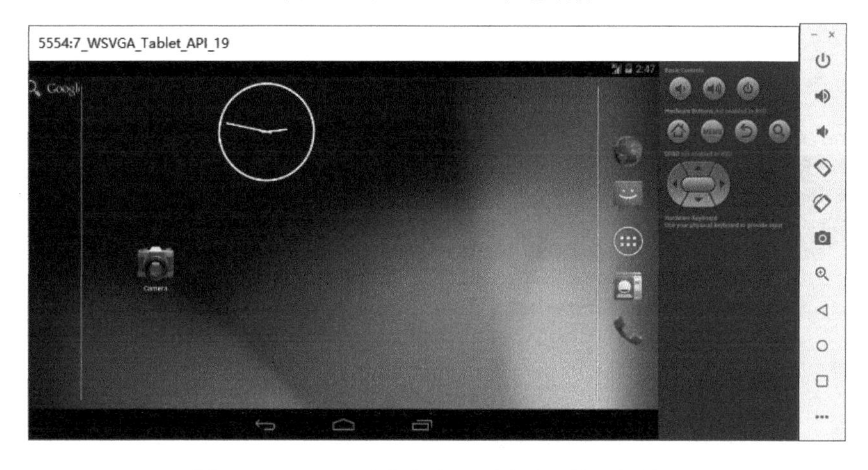

图 6.11　Android 虚拟设备配置完成后的运行效果

6.4　开发实践：Android 项目的创建与调试

6.4.1　Android 项目的创建

通过前面的学习，读者已经熟悉了 Android Studio 开发环境，接下来创建一个 Android 项目。

（1）新建项目。单击 Android Studio 开发环境的菜单“File→New Project”，会弹出如图 6.12 所示的“Create Android Project”对话框，在该对话框中填写相关信息后单击“Next”按钮。

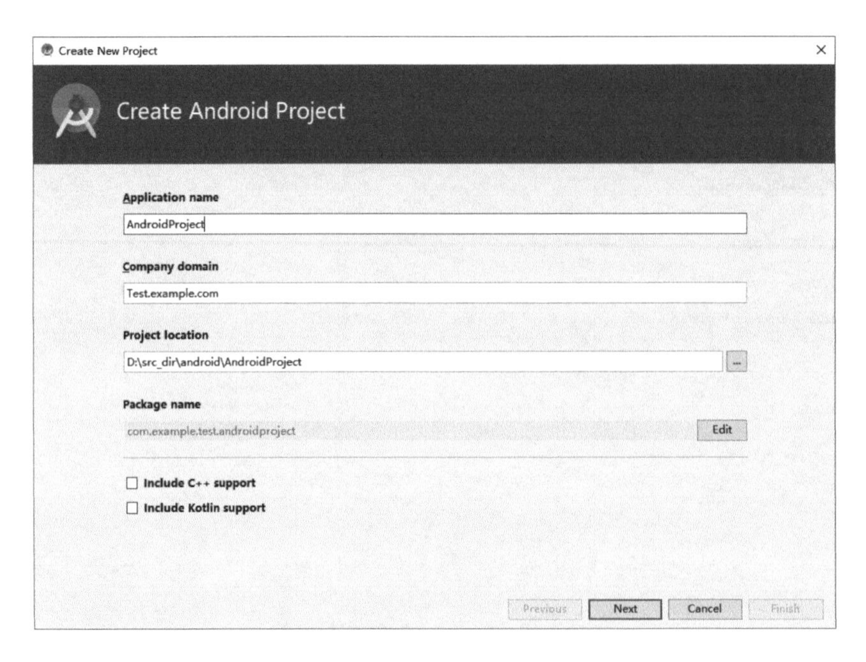

图 6.12 “Create Android Project”对话框

（2）选择 Activity 模板，如图 6.13 所示。

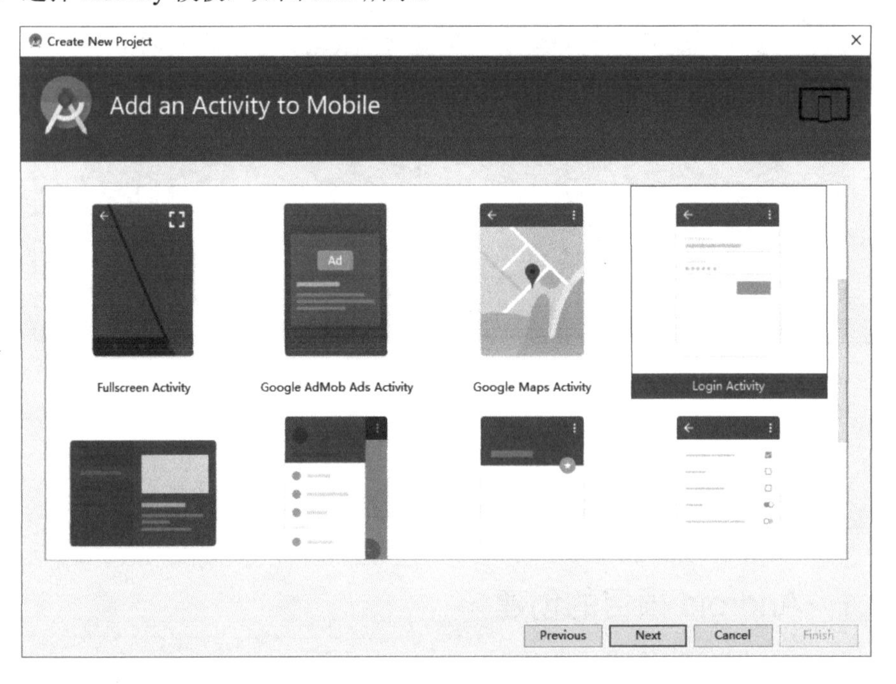

图 6.13 选择 Activity 模板

（3）完成创建项目。单击图 6.13 中的“Next”按钮会弹出“Configure Activity”对话框，如图 6.14 所示，单击“Finish”按钮即可完成 Android 项目的创建。

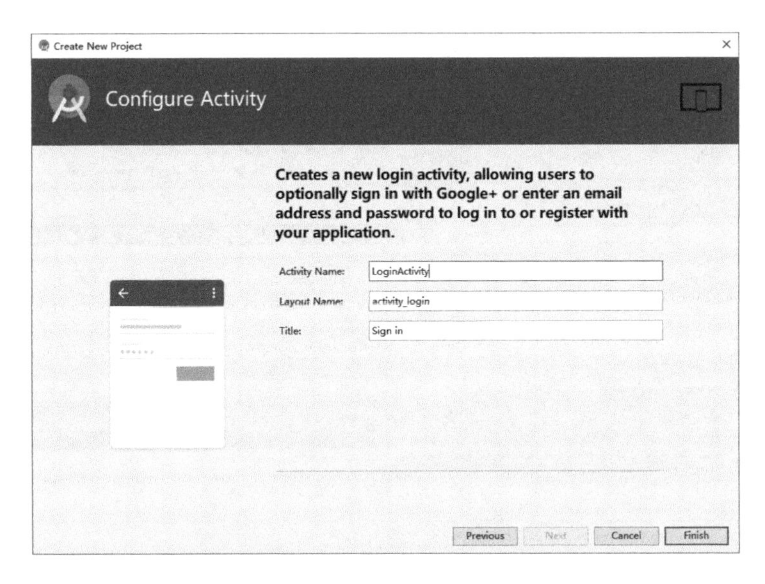

图 6.14　“Configure Activity”对话框

6.4.2　Android 项目的调试

创建 Android 项目后需要调试该项目。

（1）设置断点。在 Android Studio 开发环境中单击“app→Java→LoginActivity”，即可在打开的项目文件中设置断点。例如，双击 onCreate()方法的第 71 行即可在第 71 行设置断点。

（2）调试配置。单击 Debug 按钮即可启动项目调试。

（3）选择模拟器，如图 6.15 所示。

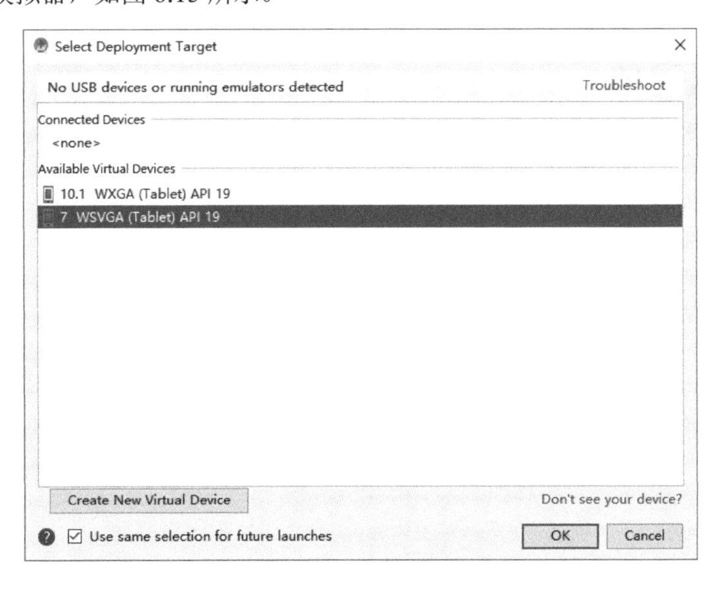

图 6.15　选择模拟器

（4）进入调试界面，Android 项目的调试界面如图 6.16 所示。

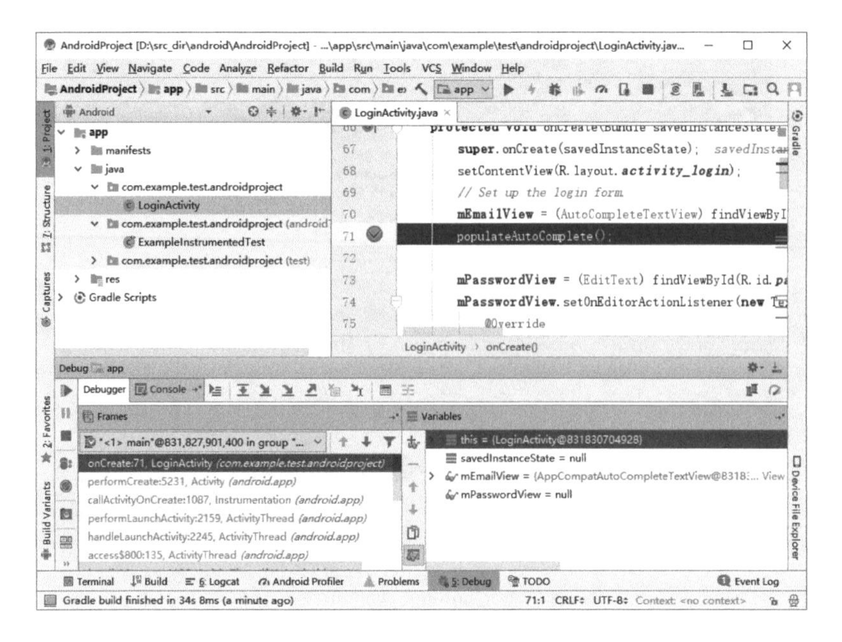

图 6.16 Android 项目的调试界面

6.5 任务验证

读者可以自己动手创建一个 Android 项目，项目创建成功后设置断点并启动项目调试。启动项目调试后单步运行程序，查看变量的值。

6.6 开发小结

本任务主要介绍了 Android 项目框架和调试。通过本任务的学习，读者可以掌握 Android 项目创建和调试的方法。

6.7 思考与拓展

（1）Android Studio 开发环境中 Project 和 Module 是什么关系？

（2）尝试单步运行程序，并查看变量的值。

任务 7

空气质量显示界面的设计

本任务主要介绍 Android 用户界面布局，主要内容包括 Android 用户界面框架、视图树、线性布局、帧布局、表格布局、相对布局和绝对布局。通过本任务的学习，读者可以掌握 Android 用户界面布局的方法，完成空气质量显示界面的设计。

7.1 开发场景：如何实现空气质量显示界面的设计

假设有一个智能空气质量监测设备，能够从该设备获取空气质量数据，如 PM1.0、PM2.5、PM10 的数值。本任务将设计空气质量显示界面，如图 7.1 所示。

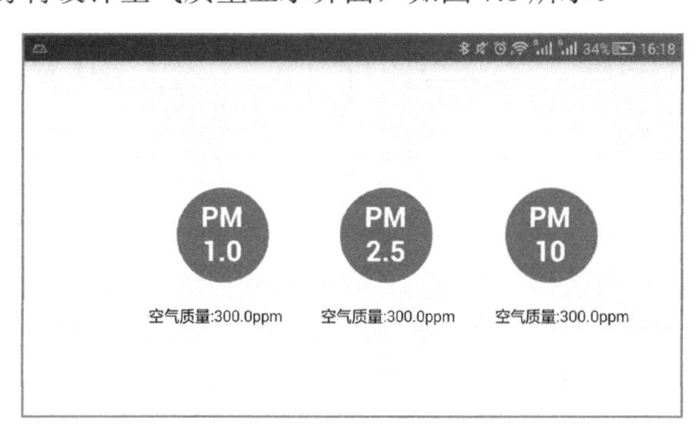

图 7.1 空气质量显示界面

7.2 开发目标

（1）知识目标：熟悉 Android 用户界面框架、视图树、线性布局、帧布局、表格布局、相对布局与绝对布局。

（2）技能目标：掌握 Android 常用布局的使用方法。

（3）任务目标：了解 Android 用户界面框架及视图树，学习 Android 的常用布局方法，实现空气质量显示界面的设计。

7.3 原理学习：Android用户界面布局

7.3.1 Android的用户界面框架

Android的用户界面通常包含活动（Activity）、片段（Fragment）、布局（Layout）、部件（Widget）等部分。Android用户界面结构如图7.2所示。

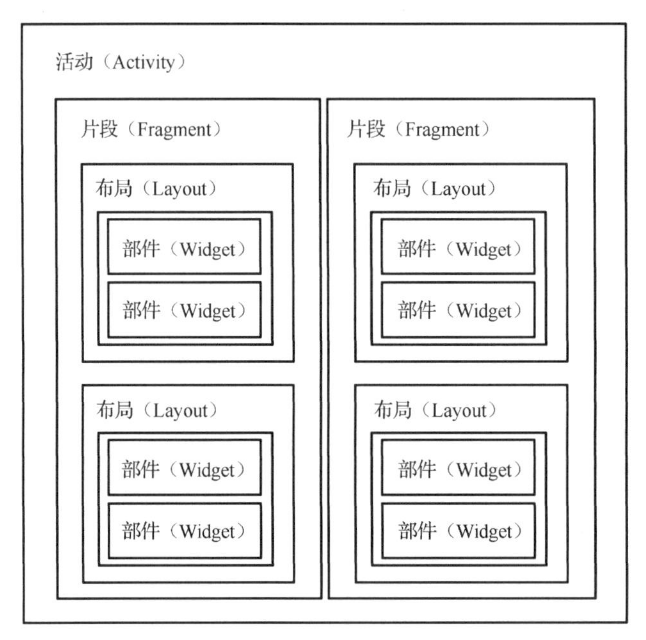

图7.2 Android用户界面结构

Activity是Android应用的核心组件，通常用来表示一个界面，也可以将多个界面添加到一个Activity中。Activity是负责界面显示内容的控件实例，既可以移除或添加内容，也可以通过Intent来启动触发新的Activity。

Fragment是界面上独立的一个部分，既可以和其他Fragment放在一起，也可以单独放置，通常把它作为一个子Activity。

Layout是对用户界面中部件进行排列设置的容器。

Widget是Android中独立的部件，包含按钮、文本框、编辑框、选择框等。

7.3.2 Android的视图树

Android用户界面框架（Android UI Framework）采用视图树（View Tree）模型，即在Android用户界面框架中的界面元素以一种树状结构组织在一起。视图树由View和ViewGroup构成，如图7.3所示。

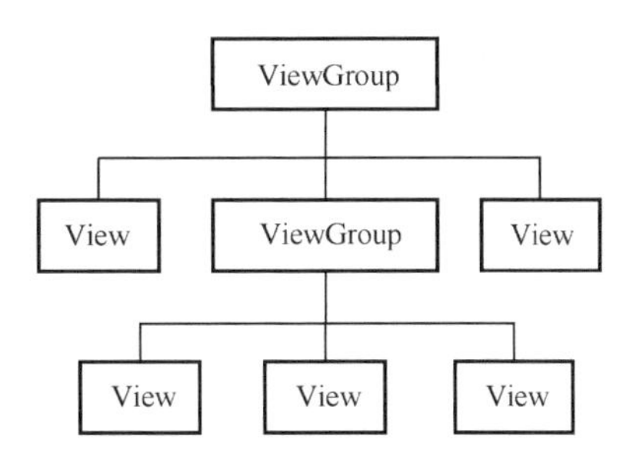

图 7.3　Android 视图树模型

Android 系统会依据视图树的结构从上至下绘制每一个界面元素。每个元素负责对自身的绘制，如果元素包含子元素，则该元素会通知其下所有子元素进行绘制。

7.3.3　Android 的线性布局

线性布局（LinearLayout）是一种常用的界面布局，也是 RadioGroup、TabWidget、TableLayout、TableRow、ZoomControls 等类的父类。在线性布局中，可以将其中的子元素以竖直或水平的方式排成一行（默认为竖直方向），通过将 android.orientation 属性设置为 horizontal（水平）或 vertical（竖直）可以达到设置线性布局的目的，线性布局不支持元素的自动浮动。线性布局示例如图 7.4 所示。

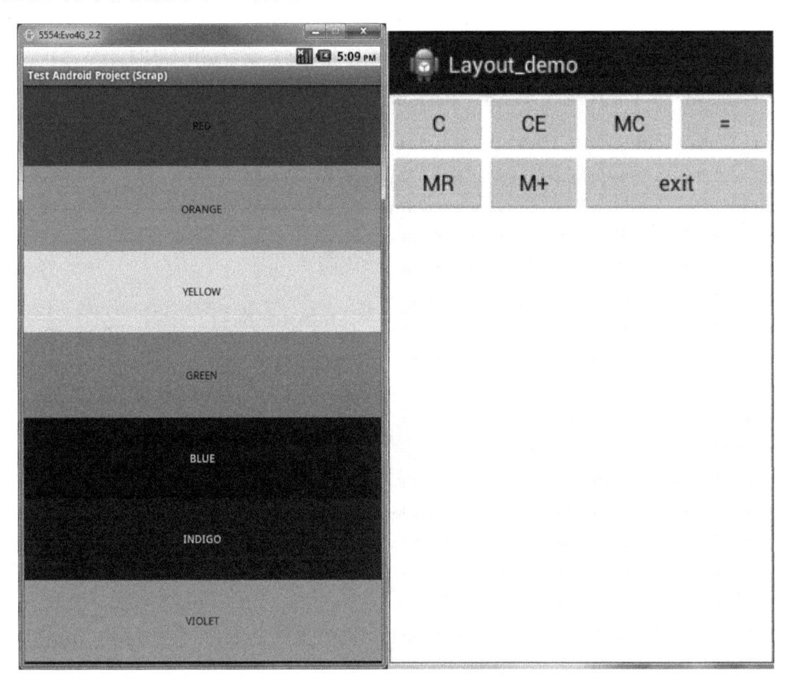

图 7.4　线性布局示例

1．常用属性

线性布局的常用属性如表 7.1 所示。

表 7.1 线性布局的常用属性

属　性	说　明
orientation	线性布局中组件的排列方式，有 horizontal（水平）和 vertical（竖直）两种方式
gravity	控制组件所包含的子元素的对齐方式，可以多个组合
layout_width	线性布局的宽度，通常不会直接设置，而是根据 wrap_content（组件的实际大小），以 fill_parent 或者 match_parent 的方式来填满父容器
layout_height	线性布局的高度，原理同 layout_width
background	为该组件设置一个背景图片或者设置背景颜色

2. Weight（权重）

线性布局根据 layout_width 和 layout_weight 来决定应该分配多少空间，遵循以下规则：

（1）根据 layout_width 的值来初步指定空间，因为 layout_width 是 wrap_content，那么布局管理器会分别给两个子控件足够的空间，用于水平方向的拉伸，如图 7.5 所示。

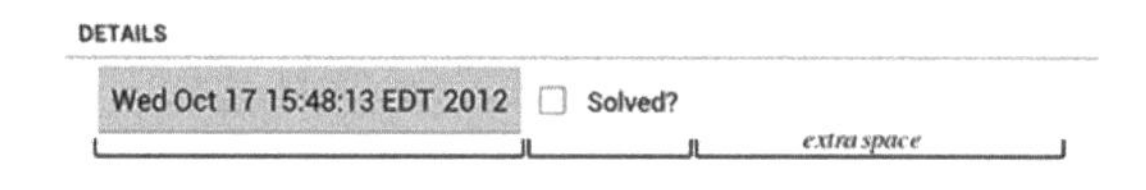

图 7.5 权重布局示例（一）

（2）如果水平方向上仍然有足够的空间，那么布局管理器就会将这个多余的控件按照 layout_weight 的值进行分配。图 7.5 所示的例子中 layout_weight 的值是 1∶1，因此布局管理器会将多余控件按照 1∶1 的比例分配，其效果如图 7.6 所示。

图 7.6 权重布局示例（二）

（3）如果想给两个控件设置相同的宽度，则 Android 推荐的做法是将两个控件的属性设置为：layout_width=0dp 及 layout_weight=1。

（4）如果一个控件没提供 layout_weight，则 Android 会默认将其设置为 0。

（5）Android 允许 layout_weight 的值是小数或整数，分配比例的规则是某个 layout_weight 的值除以所有 layout_weight 的值之和，这条规则成立的前提是线性布局没有设置 android:weightSum 来限制 layout_weight 的值的总和。

3. Divider（分割线）

（1）属性 Divider 用于线性布局设置图片的分割线，例如，“android:divider="@drawable/line_div"”。

（2）属性 showDividers 用于设置分割线所在的位置，有四个可选值，即 none、middle、beginning、end。

（3）属性 dividerPadding 用于设置分割线的填充方式。

示例代码如下：

```
<?xml version="1.0" encoding="utf-8"?>
<LinearLayout xmlns:android="http://schemas.android.com/apk/res/android"
    xmlns:tools="http://schemas.android.com/tools"
    android:layout_width="match_parent"
    android:layout_height="match_parent"
    android:orientation="vertical">
    <LinearLayout
        android:layout_width="match_parent"
        android:layout_height="match_parent"
        android:layout_weight="1"
        android:orientation="horizontal">
        <TextView
            android:layout_width="wrap_content"
            android:layout_height="match_parent"
            android:layout_weight="1"
            android:background="#aa0000"
            android:gravity="center"
            android:text="第一列"
            android:textColor="@android:color/white"
            android:textSize="20sp"></TextView>
        <TextView
            android:layout_width="wrap_content"
            android:layout_height="match_parent"
            android:layout_weight="1"
            android:background="#00aa00"
            android:gravity="center"
            android:text="第二列"
            android:textColor="@android:color/white"
            android:textSize="20sp"></TextView>
        <TextView
            android:layout_width="wrap_content"
            android:layout_height="match_parent"
            android:layout_weight="1"
            android:background="#0000aa"
            android:gravity="center"
            android:text="第三列"
            android:textColor="@android:color/white"
            android:textSize="20sp"></TextView>
        <TextView
            android:layout_width="wrap_content"
            android:layout_height="match_parent"
            android:layout_weight="1"
            android:background="#aaaa00"
            android:gravity="center"
            android:text="第四列"
            android:textColor="@android:color/white"
            android:textSize="20sp"></TextView>
```

```
    </LinearLayout>
    <LinearLayout
        android:layout_width="match_parent"
        android:layout_height="match_parent"
        android:layout_weight="1"
        android:orientation="vertical">
        <TextView
            android:layout_width="match_parent"
            android:layout_height="match_parent"
            android:layout_weight="1"
            android:background="#9C27B0"
            android:gravity="center"
            android:text="第一行"
            android:textColor="@android:color/white"
            android:textSize="20sp"></TextView>
        <TextView
            android:layout_width="match_parent"
            android:layout_height="match_parent"
            android:layout_weight="1"
            android:background="#4CAF50"
            android:gravity="center"
            android:text="第二行"
            android:textColor="@android:color/white"
            android:textSize="20sp"></TextView>
        <TextView
            android:layout_width="match_parent"
            android:layout_height="match_parent"
            android:layout_weight="1"
            android:background="#00BCD4"
            android:gravity="center"
            android:text="第三行"
            android:textColor="@android:color/white"
            android:textSize="20sp"></TextView>
        <TextView
            android:layout_width="match_parent"
            android:layout_height="match_parent"
            android:layout_weight="1"
            android:background="#FF5722"
            android:gravity="center"
            android:text="第四行"
            android:textColor="@android:color/white"
            android:textSize="20sp"></TextView>
    </LinearLayout>
</LinearLayout>
```

分割线的使用示例如图 7.7 所示。

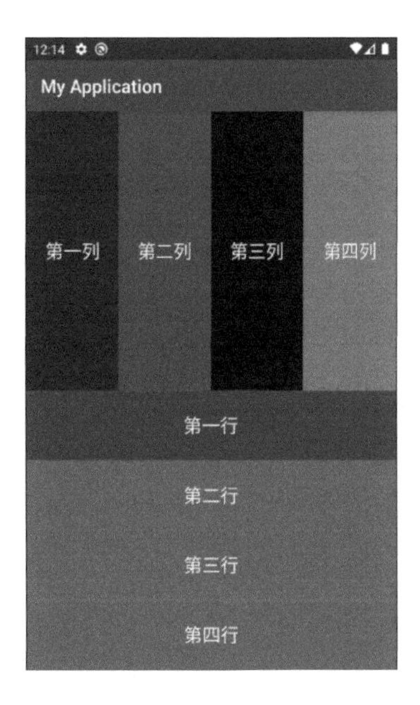

图 7.7　分割线的使用示例

7.3.4　Android 的帧布局

帧布局（FrameLayout）是最简单的布局方式，所有添加到这个布局中的视图都是以层叠的方式来显示的。第一个添加的视图放在底层，最后一个添加的视图放在顶层，上一层的视图会覆盖下一层的视图，帧布局类似堆栈布局。帧布局的属性如表 7.2 所示。

表 7.2　帧布局的属性

属　性	说　明	属　性	说　明
top	将视图放到屏幕的顶端	bottom	将视图放到屏幕的底部
left	将视图放到屏幕的左侧	right	将视图放到屏幕的右侧
center_vertical	将视图按照竖直方向居中显示	horizontal_vertical	将视图按照水平方向居中显示

示例代码如下：

```
<?xml version="1.0" encoding="utf-8"?>
<FrameLayout xmlns:android="http://schemas.android.com/apk/res/android"
    xmlns:app="http://schemas.android.com/apk/res-auto"
    xmlns:tools="http://schemas.android.com/tools"
    android:layout_width="match_parent"
    android:layout_height="match_parent"
    tools:context=".MainActivity">
    <TextView
        android:layout_width="300dp"
        android:layout_height="300dp"
        android:background="#F44336"
        android:gravity="bottom|center_horizontal"
```

```
        android:text="3"
        android:textColor="@android:color/white"
        android:textSize="25sp" />
    <TextView
        android:layout_width="200dp"
        android:layout_height="200dp"
        android:background="#4CAF50"
        android:gravity="bottom|center_horizontal"
        android:text="2"
        android:textColor="@android:color/white"
        android:textSize="25sp" />
    <TextView
        android:layout_width="100dp"
        android:layout_height="100dp"
        android:background="#FFEB3B"
        android:gravity="center"
        android:text="1"
        android:textColor="@android:color/white"
        android:textSize="25sp" />
</FrameLayout>
```

帧布局的使用示例如图 7.8 所示。

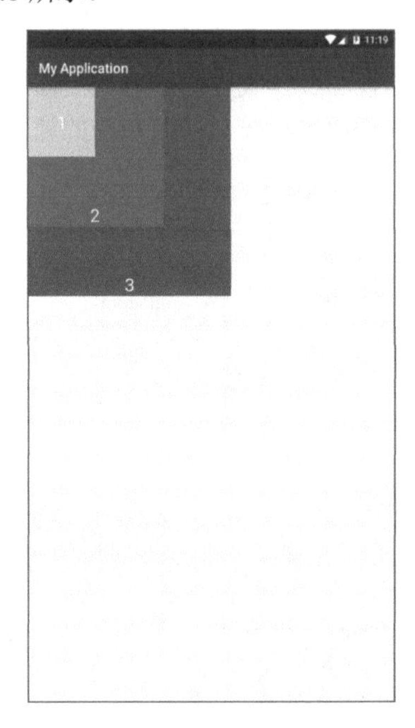

图 7.8　帧布局的使用示例

7.3.5　Android 的表格布局

表格布局（TableLayout）适用于 *N* 行 *M* 列的布局格式。一个表格布局由许多 TableRow

组成，一个 TableRow 就代表表格布局中的一行。TableRow 是线性布局的子类，它的属性值恒为 horizontal，并且它的 android:layout_width 和 android:layout_height 属性值恒为 MATCH_PARENT 和 WRAP_CONTENT。所以 TableRow 的子元素都是横向排列的，并且宽高一致。在 TableRow 中，单元格可以为空，但是不能跨列。

Android 的表格布局以行和列的形式对控件进行管理，每行都是一个 TableRow 对象或一个 View 控件。当 TableRow 是 TableRow 对象时，可在 TableRow 下添加子控件，在默认情况下，每个子控件占据一列，有多少个子控件就有多少列；当 TableRow 是 View 控件时，该 View 将独占一行。

表格布局的属性如表 7.3 所示。

表 7.3　表格布局的属性

属　性	说　明
android:layout_colum	指定该单元格在第几列显示
android:layout_span	指定该单元格占据的列数（未指定时，为 1）
android:stretchColumns	设置可拉伸的列，该列可向行方向拉伸，最多可占据一整行
android:shrinkColumns	设置可收缩的列，当该列子控件的内容太多，已经挤满所在行时，该子控件的内容将往列方向拉伸
android:layout_colum	指定该单元格在第几列显示
android:layout_span	指定该单元格占据的列数（未指定时，为 1）

1．如何确定行数与列数

（1）如果直接往表格布局中添加控件，则这个控件将占满一行。

（2）如果想要在一行上放置多个控件，就要添加一个 TableRow，把控件都放在 TableRow 中。

（3）TableRow 中的控件个数决定了该行有多少列，而列的宽度由该列中最宽的单元格决定。

（4）在默认的情况下，TableRow 的 layout_width 属性是 fill_parent，设置成其他值时也不会生效；但 layout_height 属性默认是 wrapten_content，可以设置大小。

（5）整个表格布局的宽度取决于父容器的宽度。

（6）一个 TableRow 占一行，一个单独的控件也占一行。有多少列则要看 TableRow 中的控件个数，控件最多的 TableRow 的列数就是表格布局的列数。

2．常用属性

（1）android:collapseColumns：设置需要被隐藏的列的序号。

（2）android:shrinkColumns：设置允许被收缩的列的序号。

（3）android:stretchColumns：设置运行被拉伸的列的序号。

以上三个属性的列号都是从 0 开始计算的，例如“shrinkColumns = "3"”，对应的是第 4 列，可以设置多个列，列之间用逗号隔开，如“1,3”，用“*”号可设置所有列都生效。

（4）android:layout_colum="3"：表示跳过第 3 个单元格，直接在第 4 个单元格处显示，序号从 1 开始计算。

（5）android:layout_span="5"：表示合并 5 个单元格，也就说这个控件占 5 个单元格。

简单的表格布局示例代码如下：

```
<?xml version="1.0" encoding="utf-8"?>
<TableLayout xmlns:android="http://schemas.android.com/apk/res/android"
    xmlns:app="http://schemas.android.com/apk/res-auto"
    xmlns:tools="http://schemas.android.com/tools"
    android:layout_width="match_parent"
    android:layout_height="match_parent"
    tools:context=".MainActivity">
    <TableRow>
        <TextView
            android:layout_width="100dp"
            android:layout_height="100dp"
            android:background="#F44336"
            android:gravity="center"
            android:text="1"
            android:textColor="@android:color/white"
            android:textSize="25sp" />
        <TextView
            android:layout_width="100dp"
            android:layout_height="100dp"
            android:background="#673AB7"
            android:gravity="center"
            android:text="2"
            android:textColor="@android:color/white"
            android:textSize="25sp" />
        <TextView
            android:layout_width="100dp"
            android:layout_height="100dp"
            android:background="#2196F3"
            android:gravity="center"
            android:text="3"
            android:textColor="@android:color/white"
            android:textSize="25sp" />
    </TableRow>
    <TableRow>
        <TextView
            android:layout_width="100dp"
            android:layout_height="100dp"
            android:background="#009688"
            android:gravity="center"
            android:text="4"
            android:textColor="@android:color/white"
            android:textSize="25sp" />
        <TextView
            android:layout_width="100dp"
```

```
            android:layout_height="100dp"
            android:background="#CDDC39"
            android:gravity="center"
            android:text="5"
            android:textColor="@android:color/white"
            android:textSize="25sp" />
        <TextView
            android:layout_width="100dp"
            android:layout_height="100dp"
            android:background="#FFC107"
            android:gravity="center"
            android:text="6"
            android:textColor="@android:color/white"
            android:textSize="25sp" />
    </TableRow>
    <TableRow>
        <TextView
            android:layout_width="100dp"
            android:layout_height="100dp"
            android:background="#FF9800"
            android:gravity="center"
            android:text="7"
            android:textColor="@android:color/white"
            android:textSize="25sp" />
        <TextView
            android:layout_width="100dp"
            android:layout_height="100dp"
            android:background="#9C27B0"
            android:gravity="center"
            android:text="8"
            android:textColor="@android:color/white"
            android:textSize="25sp" />
        <TextView
            android:layout_width="100dp"
            android:layout_height="100dp"
            android:background="#4CAF50"
            android:gravity="center"
            android:text="9"
            android:textColor="@android:color/white"
            android:textSize="25sp" />
    </TableRow>
</TableLayout>
```

表格布局的使用示例如图 7.9 所示。

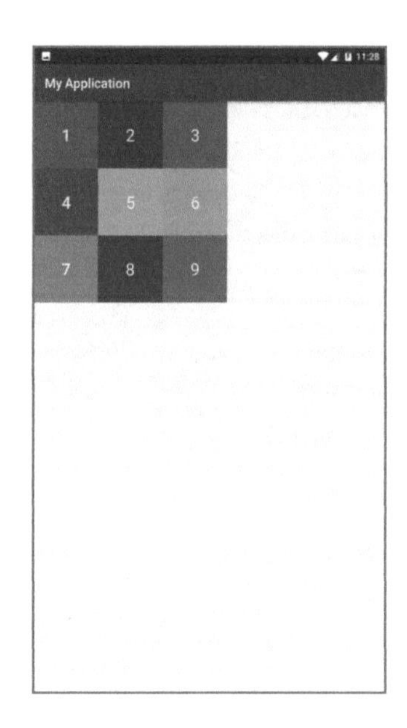

图 7.9　表格布局的使用示例

7.3.6　Android 的相对布局

相对布局（RelativeLayout）是以兄弟控件或者父容器来决定控件的显示位置的。合理地利用好线性布局的 Weight 属性和相对布局，可以解决不同屏幕分辨率的自适应问题。相对布局界面示例如图 7.10 所示。

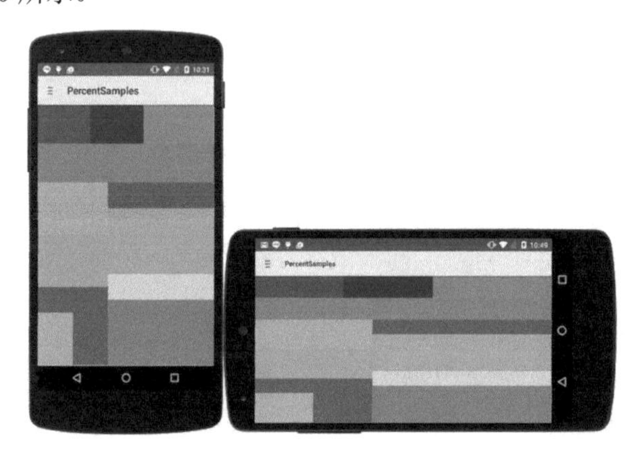

图 7.10　相对布局界面示例

1．对齐方式

android:gravity 用于设置容器内各个子控件的对齐方式。如果将哪个控件设置为 android:ignoreGravity，则该控件不受 Gravity 属性的影响。

（1）根据父容器来定位。

① android:layout_alighParentLeft：左对齐。

② android:layout_alighParentRight：右对齐。

③ android:layout_alighParentTop：顶端对齐。
④ android:layout_alighParentBottom：底部对齐。
⑤ android:layout_centerHorizontal：水平居中。
⑥ android:layout_centerVertical：竖直居中。
⑦ android:layout_centerInParent：中央位置。
（2）根据兄弟控件来定位。
① android:layout_toLeftOf：左对齐。
② android:layout_toRightOf：右对齐。
③ android:layout_above：上对齐。
④ android:layout_below：下对齐。
⑤ android:layout_alignTop：上边界对齐。
⑥ android:layout_alignBottom：下边界对齐。
⑦ android:layout_alignLeft：左边界对齐。
⑧ android:layout_alignRight：右边界对齐。

2. Margin 和 Padding

（1）Margin 用于设置控件与父容器（通常是布局）的边距。
① android:layout_margin：控件的四周外部留出一定的边距。
② android:layout_marginLeft：控件的左边外部留出一定的边距。
③ android:layout_marginTop：控件的上边外部留出一定的边距。
④ android:layout_marginRight：控件的右边外部留出一定的边距。
⑤ android:layout_marginBottom：控件的下边外部留出一定的边距。
（2）Padding 用于设置控件内部元素间的边距（可以理解为填充）。
① android:padding：控件的四周的内部留出一定的边距。
② android:paddingLeft：控件的左边的内部留出一定的边距。
③ android:paddingTop：控件的上边的内部留出一定的边距。
④ android:paddingRight：控件的右边的内部留出一定的边距。
⑤ android:paddingBottom：控件的下边的内部留出一定的边距。

3. 其他属性

① android:gravity 用来对 View 控件的内容进行限定。例如，可以设置 Button 中文本的位置。以 Button 为例，“android:gravity="right"”表示 Button 上面的文本靠右。

② android:layout_gravity 用来设置 View 控件相对于其父 View 控件的位置。例如，在线性布局里，如果想把 Button 放在靠左、靠右等位置，就可以通过该属性来设置。以 Button 为例，“android:layout_gravity="right"”表示 Button 放在靠右的位置。

③ android:layout_alignParentRight 用于设置当前控件的右端和其父控件的右端对齐，这里属性值只能为 true 或 false，默认为 false。

④ android:scaleType 用来调整图片以匹配 ImageView 的大小。

示例代码如下：

```
<?xml version="1.0" encoding="utf-8"?>
<RelativeLayout xmlns:android="http://schemas.android.com/apk/res/android"
    xmlns:app="http://schemas.android.com/apk/res-auto"
    xmlns:tools="http://schemas.android.com/tools"
    android:layout_width="match_parent"
    android:layout_height="match_parent"
    tools:context=".MainActivity">
    <TextView
        android:layout_width="100dp"
        android:layout_height="100dp"
        android:background="#F44336"
        android:gravity="center"
        android:text="1"
        android:textColor="@android:color/white"
        android:textSize="25sp" />
    <TextView
        android:layout_width="100dp"
        android:layout_height="100dp"
        android:layout_alignParentRight="true"
        android:background="#673AB7"
        android:gravity="center"
        android:text="2"
        android:textColor="@android:color/white"
        android:textSize="25sp" />
    <TextView
        android:layout_width="100dp"
        android:layout_height="100dp"
        android:layout_alignParentBottom="true"
        android:background="#2196F3"
        android:gravity="center"
        android:text="3"
        android:textColor="@android:color/white"
        android:textSize="25sp" />
    <TextView
        android:layout_width="100dp"
        android:layout_height="100dp"
        android:layout_alignParentRight="true"
        android:layout_alignParentBottom="true"
        android:background="#009688"
        android:gravity="center"
        android:text="4"
        android:textColor="@android:color/white"
        android:textSize="25sp" />
    <TextView
        android:layout_width="100dp"
        android:layout_height="100dp"
        android:layout_centerInParent="true"
```

```
        android:background="#CDDC39"
        android:gravity="center"
        android:text="5"
        android:textColor="@android:color/white"
        android:textSize="25sp" />
    <TextView
        android:layout_width="100dp"
        android:layout_height="100dp"
        android:layout_centerVertical="true"
        android:background="#FFC107"
        android:gravity="center"
        android:text="6"
        android:textColor="@android:color/white"
        android:textSize="25sp" />
    <TextView
        android:layout_width="100dp"
        android:layout_height="100dp"
        android:layout_centerHorizontal="true"
        android:background="#FF9800"
        android:gravity="center"
        android:text="7"
        android:textColor="@android:color/white"
        android:textSize="25sp" />
    <TextView
        android:layout_width="100dp"
        android:layout_height="100dp"
        android:layout_alignParentRight="true"
        android:layout_centerVertical="true"
        android:background="#9C27B0"
        android:gravity="center"
        android:text="8"
        android:textColor="@android:color/white"
        android:textSize="25sp" />
    <TextView
        android:layout_width="100dp"
        android:layout_height="100dp"
        android:layout_alignParentBottom="true"
        android:layout_centerHorizontal="true"
        android:background="#4CAF50"
        android:gravity="center"
        android:text="9"
        android:textColor="@android:color/white"
        android:textSize="25sp" />
</RelativeLayout>
```

相对布局的使用示例如图 7.11 所示。

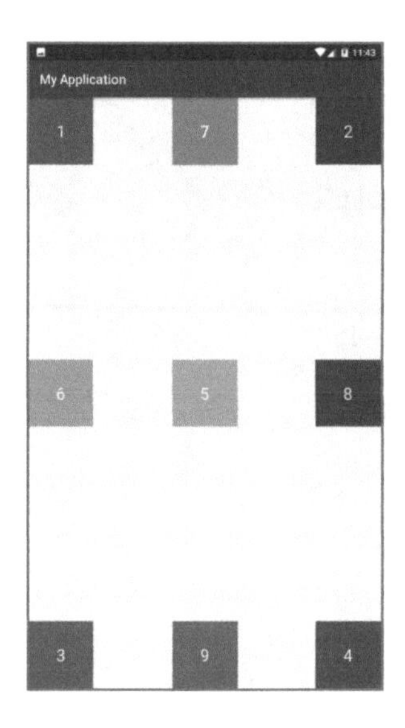

图 7.11　相对布局的使用示例

7.3.7　Android 的绝对布局

绝对布局（AbsoluteLayout）的常用属性如下：

（1）android:layout_width：用于设置控件宽度。

（2）android:layout_height：用于设置控件高度。

（3）android:layout_x：用于设置控件水平方向的坐标。

（4）android:layout_y：用于设置控件竖直方向的坐标。

示例代码如下：

```
<?xml version="1.0" encoding="utf-8"?>
<AbsoluteLayout xmlns:android="http://schemas.android.com/apk/res/android"
    xmlns:app="http://schemas.android.com/apk/res-auto"
    xmlns:tools="http://schemas.android.com/tools"
    android:layout_width="match_parent"
    android:layout_height="match_parent"
    tools:context=".MainActivity">
    <TextView
        android:layout_width="100dp"
        android:layout_height="100dp"
        android:background="#F44336"
        android:gravity="center"
        android:text="1"
        android:textColor="@android:color/white"
        android:textSize="25sp" />
    <TextView
        android:layout_width="100dp"
        android:layout_height="100dp"
        android:layout_x="200dp"
```

```
        android:layout_y="0dp"
        android:background="#673AB7"
        android:gravity="center"
        android:text="2"
        android:textColor="@android:color/white"
        android:textSize="25sp" />
    <TextView
        android:layout_width="100dp"
        android:layout_height="100dp"
        android:layout_x="0dp"
        android:layout_y="200dp"
        android:background="#2196F3"
        android:gravity="center"
        android:text="3"
        android:textColor="@android:color/white"
        android:textSize="25sp" />
    <TextView
        android:layout_width="100dp"
        android:layout_height="100dp"
        android:layout_x="200dp"
        android:layout_y="200dp"
        android:background="#009688"
        android:gravity="center"
        android:text="4"
        android:textColor="@android:color/white"
        android:textSize="25sp" />
</AbsoluteLayout>
```

绝对布局的使用示例如图 7.12 所示。

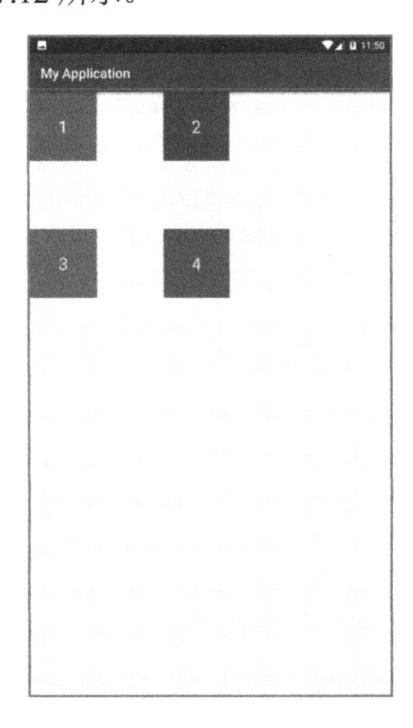

图 7.12　绝对布局的使用示例

7.4 开发实践：空气质量显示界面

7.4.1 开发设计

本任务要实现的功能是在空气质量显示界面中显示 PM1.0、PM2.5 和 PM10 三种数据，采用相对布局和线性布局来实现。项目布局框架如图 7.13 所示。

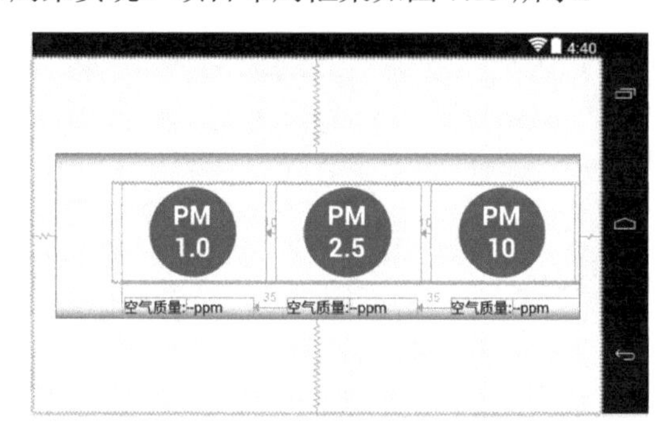

图 7.13 项目布局框架

本项目的布局框架采用三层布局设计：最外层采用相对布局；中间一层采用一个线性布局，设置为竖直方式；线性布局下面有两个相对布局，分别是图标与文字说明。项目布局结构如图 7.14 所示。

图 7.14 项目布局结构

7.4.2 功能实现

1. 布局设计

定义布局文件 activity_main.xml，代码如下：

```
<?xml version="1.0" encoding="utf-8"?>
<RelativeLayout xmlns:android="http://schemas.android.com/apk/res/android"
    android:layout_width="fill_parent"
```

```
android:layout_height="fill_parent">
<LinearLayout
    android:layout_width="wrap_content"
    android:layout_height="wrap_content"
    android:layout_gravity="center"
    android:orientation="vertical"
    android:layout_centerInParent="true">
    <RelativeLayout
        android:id="@+id/edit_div"
        android:layout_width="fill_parent"
        android:layout_height="wrap_content"
        android:layout_marginLeft="60dp"
        android:layout_marginTop="30dp">
        <ImageView
            android:id="@+id/air_image2"
            android:layout_width="150dp"
            android:layout_height="100dp"
            android:layout_gravity="center"
            android:layout_marginLeft="10dp"
            android:layout_toRightOf="@+id/air_image1"
            android:src="@drawable/pm25"/>
        <ImageView
            android:id="@+id/air_image3"
            android:layout_width="150dp"
            android:layout_height="100dp"
            android:layout_gravity="center"
            android:layout_marginLeft="10dp"
            android:layout_toRightOf="@+id/air_image2"
            android:src="@drawable/pm100"/>
    </RelativeLayout>
    <RelativeLayout
        android:id="@+id/edit_div2"
        android:layout_width="fill_parent"
        android:layout_height="wrap_content"
        android:layout_marginLeft="70dp">
        <TextView
            android:id="@+id/air_label1"
            android:layout_width="wrap_content"
            android:layout_height="wrap_content"
            android:layout_gravity="center"
            android:layout_marginLeft="3dp"
            android:text="空气质量:"
            android:textColor="@color/black"
            android:layout_marginTop="15dp"
            android:textSize="15.0sp"/>
        <TextView
            android:id="@+id/air_tv1"
```

```
        android:layout_width="70dp"
        android:layout_height="wrap_content"
        android:layout_gravity="center"
        android:layout_toRightOf="@+id/air_label1"
        android:text="--ppm"
        android:textColor="@color/black"
        android:layout_marginTop="15dp"
        android:textSize="15.0sp"/>
    <TextView
        android:id="@+id/air_label2"
        android:layout_width="wrap_content"
        android:layout_height="wrap_content"
        android:layout_gravity="center"
        android:layout_marginLeft="35dp"
        android:layout_toRightOf="@+id/air_tv1"
        android:text="空气质量:"
        android:textColor="@color/black"
        android:layout_marginTop="15dp"
        android:textSize="15.0sp"/>
    <TextView
        android:id="@+id/air_tv2"
        android:layout_width="70dp"
        android:layout_height="wrap_content"
        android:layout_gravity="center"
        android:layout_toRightOf="@+id/air_label2"
        android:text="--ppm"
        android:textColor="@color/black"
        android:layout_marginTop="15dp"
        android:textSize="15.0sp"/>
    <TextView
        android:id="@+id/air_label3"
        android:layout_width="wrap_content"
        android:layout_height="wrap_content"
        android:layout_gravity="center"
        android:layout_marginLeft="35dp"
        android:layout_toRightOf="@+id/air_tv2"
        android:text="空气质量:"
        android:textColor="@color/black"
        android:layout_marginTop="15dp"
        android:textSize="15.0sp"/>
    <TextView
        android:id="@+id/air_tv3"
        android:layout_width="70dp"
        android:layout_height="wrap_content"
        android:layout_gravity="center"
        android:layout_toRightOf="@+id/air_label3"
        android:text="--ppm"
```

```
                    android:textColor="@color/black"
                    android:layout_marginTop="15dp"
                    android:textSize="15.0sp"/>
        </RelativeLayout>
    </LinearLayout>
</RelativeLayout>
```

2. 设计项目清单文件

```
<?xml version="1.0" encoding="utf-8"?>
<manifest xmlns:android="http://schemas.android.com/apk/res/android"
    package="com.example.a.myapplication">
    <application
        android:allowBackup="true"
        android:icon="@mipmap/ic_launcher"
        android:label="@string/app_name"
        android:roundIcon="@mipmap/ic_launcher_round"
        android:supportsRtl="true"
        android:theme="@style/AppTheme">
        <activity
            android:name=".MainActivity"
            android:screenOrientation="landscape">
            <intent-filter>
                <action android:name="android.intent.action.MAIN" />
                <category android:name="android.intent.category.LAUNCHER" />
            </intent-filter>
        </activity>
    </application>
</manifest>
```

3. 设计主程序

本例程的主程序入口为MainActivity，代码如下：

```
package com.example.a.myapplication;
import android.app.Activity;
import android.os.Bundle;
import android.view.Window;
import android.widget.TextView;
public class MainActivity extends Activity {
    private TextView t1;
    private TextView t2;
    private TextView t3;
    @Override
    protected void onCreate(Bundle savedInstanceState) {
        super.onCreate(savedInstanceState);
        requestWindowFeature(Window.FEATURE_NO_TITLE);
        setContentView(R.layout.activity_main);
```

```
            t1 = findViewById(R.id.air_tv1);
            t2 = findViewById(R.id.air_tv2);
            t3 = findViewById(R.id.air_tv3);
float pm1 = 300f;
            float pm2 = 300f;
            float pm3 = 300f;
            setValue(t1, pm1);
            setValue(t2, pm2);
            setValue(t3, pm3);
        }
        private void setValue(TextView textView, float pm) {
            textView.setText(pm+"ppm");
        }
}
```

7.5 任务验证

在 Android Studio 开发环境中打开本项目的例程，编译通过后运行程序。程序运行效果如图 7.15 所示。

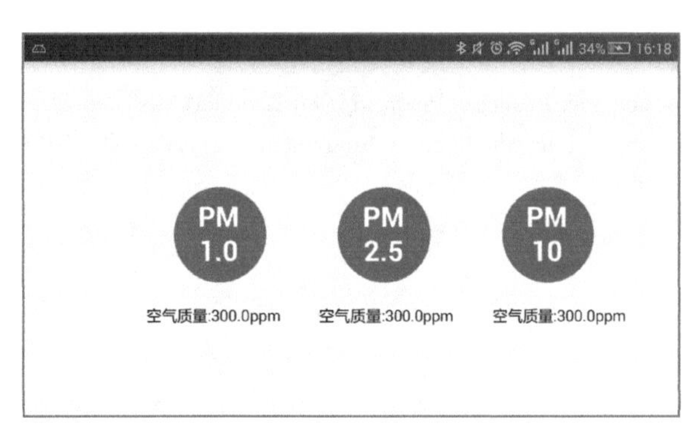

图 7.15　程序运行效果

7.6 开发小结

本任务介绍了 Android 用户界面布局的相关内容，通过空气质量显示界面的设计加深了读者对 Android 用户界面布局的理解。

7.7 思考与拓展

（1）Android 绝对布局和相对布局有什么区别？

（2）请尝试增加单击 PM 图标的响应事件，单击 PM 图标后，在屏幕弹出提示信息。

任务 8

城市气象监控设备管理系统应用界面的设计

本任务介绍 Android 界面控件的基础知识，重点介绍 TextView、EditText、Button、ImageButton、CheckBox、RadioButton、Spinner、ListView 等控件的使用方法，完成城市气象监控设备管理系统应用界面的设计。

8.1 开发场景：如何设计城市气象监控设备管理系统应用界面

在物联网中，很多物品（如手机、车辆、雨伞等）都可能成为获取气象数据的来源，尤其是随着可穿戴设备的不断发展，未来每个人都有可能成为气象数据的来源。

假如有一个气象观测站，有很多气象观测设备，该如何开发一个 Android 应用来对这些设备进行管理呢？这就是本任务需要完成的开发实践。

8.2 开发目标

（1）知识目标：熟悉 TextView、EditText、Button、ImageButton、CheckBox、RadioButton、Spinner 和 ListView 等控件。

（2）技能目标：掌握 TextView、EditText、Button、ImageButton、CheckBox、RadioButton、Spinner 和 ListView 等控件的使用方法。

（3）任务目标：通过学习 Android 界面控件的基础知识，完成城市气象监控设备管理系统应用界面的设计。

8.3 原理学习：Android 界面控件基础

8.3.1 TextView 控件

TextView 控件继承自 View 类，是一个只读文本框，可支持多行显示、字符串格式化以及自动换行等，通过 TextView 类的属性和方法可以设置 TextView 控件的显示特性。TextView

控件如图 8.1 所示。

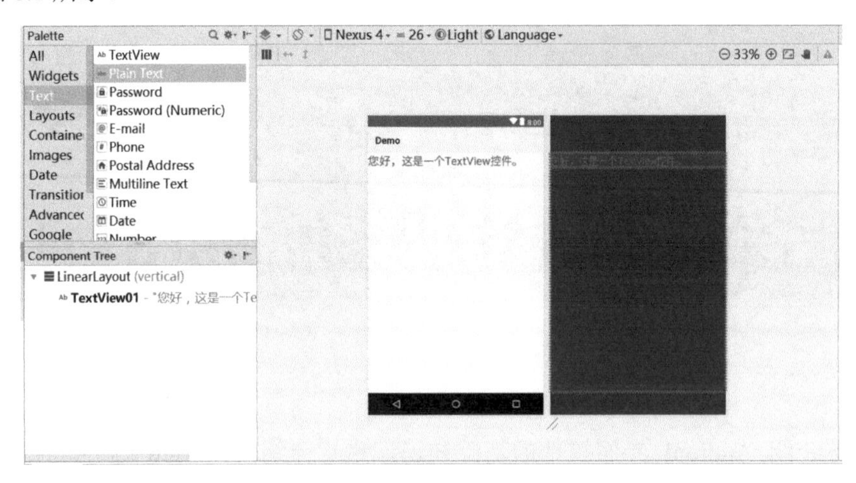

图 8.1　TextView 控件

TextView 类的部分属性和方法如表 8.1 所示。

表 8.1　TextView 类的部分属性和方法

属　性	方　法	说　明
android:text	setText(CharSquence)、setText(int resId)	设置文本框显示的文本内容
android:textColor	setTextColor(ColorStateList)	设置文本框显示的文本颜色
android:textSize	setTextSize(float)	设置文本框显示的文本字号
android:textStyle	setTypeface(Typeface)	设置文本框显示的文本字体风格
android:textAppearance	—	设置文本框显示的文本颜色、字体、大小等样式
android:ellipsize	setEllipsize(TextUitls.TruncateAt)	设置要显示的文本超出 TextView 控件的长度时，处理文本的方式
android:gravity	setGravity(int)	设置文本框显示文本的对齐方式
android:maxLines	setMaxLines(int)	设置文本框最多占几行
android:minLines	setMinLines(int)	设置文本框最少占几行
android:singleLine	setTransformationMethod	设置文本框是否是单行模式
android:drawableLeft	setCompoundDrawablesWithIntrinsicBounds (Drawable left,Drawable top,Drawable right,Drawable bottom)	在文本框左侧绘制图像

TextView 控件的示例代码如下：

```
<?xml version="1.0" encoding="utf-8"?>
<?xml version="1.0" encoding="utf-8"?>
<LinearLayout xmlns:android="http://schemas.android.com/apk/res/android"
    xmlns:app="http://schemas.android.com/apk/res-auto"
    xmlns:tools="http://schemas.android.com/tools"
    android:layout_width="match_parent"
    android:layout_height="match_parent"
```

```
    tools:context=".MainActivity">
    <TextView
        android:layout_width="100dp"
        android:layout_height="100dp"
        android:layout_marginLeft="30dp"
        android:layout_marginTop="30dp"
        android:background="#F44336"
        android:gravity="center"
        android:text="1"
        android:textColor="@android:color/white"
        android:textSize="25sp"/>
</LinearLayout>
```

TextView 控件的使用示例如图 8.2 所示。

图 8.2　TextView 控件的使用示例

8.3.2　EditText 控件

EditText 控件继承自 android.widget.TextView，是可编辑的文本框，其主要功能是输入和编辑字符串。

EditText 控件的使用方法也有两种：

（1）用 XML 描述一个 EditText 控件。

```
< EditText android:id="@+id/edit_text"
android:layout_width="fill_parent"
```

```
android:layout_height="wrap_content"
android:hint="这里可以输入文字" />
```

（2）在程序中引用 XML 描述的 EditText 控件。

```
EditText eidtText=( EditText)findViewById(R.id.editText);
```

EditText 控件的部分属性和方法如表 8.2 所示。

表 8.2　EditText 控件的部分属性和方法

属　性	方　法	说　明
android:hint	setHint(int)	文本框的提示文字
android:password	setTransformationMethod(TransformationMethod)	设置文本框中的内容是否显示为密码，当该属性为 true 时，以小黑点显示文本
android:phoneNumber	setKeyListener(KeyListner)	设置文本框中的内容只能是电话号码，当该属性为 ture 时，表示电话框
android:digits	setKeyListener(KeyListner)	设置允许输入哪些字符
android:numeric	setKeyListener(KeyListner)	设置只能输入数字，并且置顶可输入的格式，可选值有 integer、signed、decimal
android:singleLine	setTransformationMethod(TransformationMethod)	设置文本框的单行模式
android:maxLength	setFilters(InputFilter)	设置最大显示长度
android:cursorVisible	setCursorVisible(boolean)	设置光标是否可见，默认为可见

EditText 控件的示例代码如下：

```
<?xml version="1.0" encoding="utf-8"?>
<LinearLayout xmlns:android="http://schemas.android.com/apk/res/android"
    xmlns:app="http://schemas.android.com/apk/res-auto"
    xmlns:tools="http://schemas.android.com/tools"
    android:layout_width="match_parent"
    android:layout_height="match_parent"
    android:layout_margin="20dp"
    android:orientation="vertical"
    tools:context=".MainActivity">
    <EditText
        android:id="@+id/edtName"
        android:layout_width="match_parent"
        android:layout_height="wrap_content"
        android:hint="请输入您的姓名"
        android:textSize="25sp" />
    <Button
        android:id="@+id/btn_ok"
        android:layout_width="match_parent"
        android:layout_height="wrap_content"
        android:background="#4CAF50"
        android:text="确定"/>
</LinearLayout>
```

EditText 控件的示例代码如下：

```
public class MainActivity extends AppCompatActivity {
    private EditText edtName;
    private Button btnOk;
    @Override
    protected void onCreate(Bundle savedInstanceState) {
        super.onCreate(savedInstanceState);
        setContentView(R.layout.activity_main);
        edtName = findViewById(R.id.edtName);
        btnOk = findViewById(R.id.btn_ok);
        btnOk.setOnClickListener(new View.OnClickListener() {
            @Override
            public void onClick(View v) {
                String name = edtName.getText().toString();
                Toast.makeText(MainActivity.this,"您的姓名是" + name, Toast.LENGTH_SHORT).show();
            }
        });
    }
}
```

EditText 控件的使用实例如图 8.3 所示。

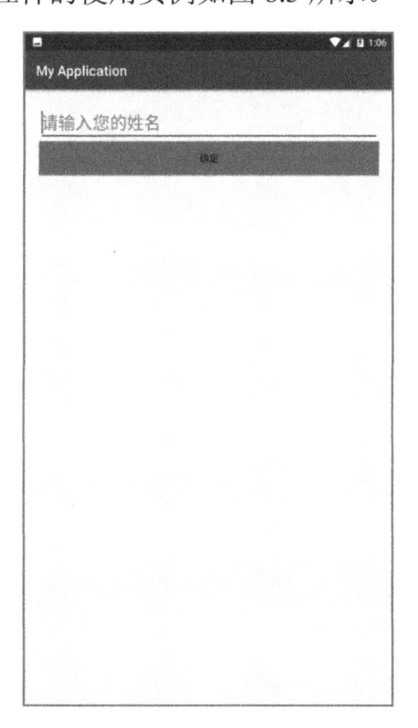

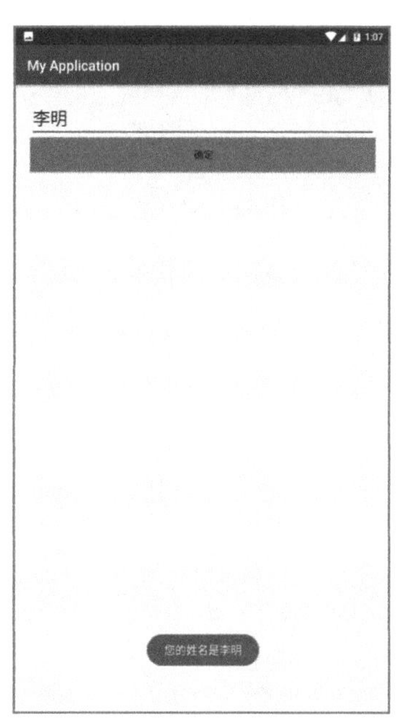

图 8.3　EditText 控件的使用实例

8.3.3　Button 控件

Button 是一种常用的按钮控件，继承自 android.widget.TextView，如图 8.4 所示。用户可以通过单击 Button 控件来触发相应的事件处理函数。

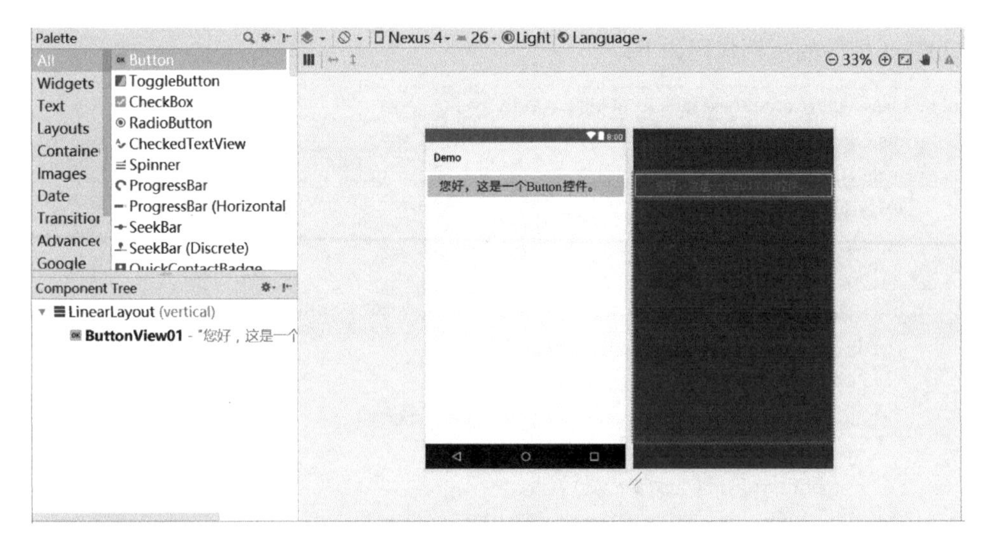

图 8.4　Button 控件

在程序中，可以通过 super.findViewById(id)来引用 XML 文件中声明的 Button 控件，然后使用 setOnClickListener(View.OnClickListener)添加监听事件，接着在 View.OnClickListener 监听器中使用 v.equals(View)方法来判断单击的是哪个控件，以此来调用不同的事件处理函数。

（1）在 XML 文件中声明一个 Button 控件，代码如下：

```
<Button Android:id="@+id/button"
Android:layout_width="wrap_content"
Android:layout_height="wrap_content"
Android:text="这是一个 Button" />
```

（2）在程序中引用 XML 文件中声明的 Button 控件，代码如下：

```
Button button = (Button) findViewById(R.id.button);
```

（3）为 Button 控件添加监听事件，代码如下：

```
button.setOnClickListener(button_listener);
```

（4）生成一个事件监听器，代码如下：

```
private Button.OnClickListener button_listener = new Button.OnClickListener(){
    public void onClick(View v){
        switch(v.getId()){
            case R.id.Button:
            textView.setText("Button 按钮 1");
            return;
            case R.id.Button01:
            textView.setText("Button 按钮 2");
            return;
        }
    }
```

```
}
```

也可以采用在 XML 文件中分配一个方法给 Button 控件，如使用 android:onClick 属性，代码如下：

```
<Button
android:layout_height=”wrap_content”
android:layout_width=”wrap_content”
android:text=”@string/self_destruct”
android:onClick=”self-destruct” />
```

当单击控件时，Android 会自动调用 Activity 中的 selfDestruct(View)方法，但 selfDestruct(View)方法必须声明为 public，并且 View 是其唯一的参数。传递 selfDestruct(View)方法的 View 是被单击控件的一个引用，代码如下：

```
public void self-destruct(View view){
    //Kabloey
}
```

（5）为 Button 控件添加事件监听器，代码如下：

```
mButton.setOnClickListener();                //单击事件监听器
mButton.setOnTouchListener();                //触摸事件监听器
mButton.setOnFocusChangeListener();          //焦点状态改变事件监听器
mButton.setOnKeyListener();                  //按键事件监听器
mButton.setOnLongClickListener();            //长按事件监听器
```

（6）控件图文混排。在 XML 文件中，可以通过设置以下几个属性来实现图片环绕文字的效果。

① android:drawableTop：用于设置在文字上方显示图片。

② android:drawableBottom：用于设置在文字下方显示图片。

③ android:drawableLeft：用于设置在文字左边显示图片。

④ android:drawableRight：用于设置在文字右边显示图片。

示例代码如下：

```
<?xml version="1.0" encoding="utf-8"?>
<LinearLayout xmlns:android="http://schemas.android.com/apk/res/android"
    xmlns:app="http://schemas.android.com/apk/res-auto"
    xmlns:tools="http://schemas.android.com/tools"
    android:layout_width="match_parent"
    android:layout_height="match_parent"
    android:layout_margin="20dp"
    android:orientation="vertical"
    tools:context=".MainActivity">
    <Button
        android:id="@+id/btn_ok"
        android:layout_width="wrap_content"
        android:layout_height="wrap_content"
        android:background="#4CAF50"
```

```
            android:text="确定" />
        <Button
            android:id="@+id/btn_cancel"
            android:layout_width="wrap_content"
            android:layout_height="wrap_content"
            android:layout_marginTop="20dp"
            android:background="#F44336"
            android:text="取消" />
</LinearLayout>
public class MainActivity extends AppCompatActivity implements View.OnClickListener{
        private Button btnOk;
        private Button btnCancel;
        @Override
        protected void onCreate(Bundle savedInstanceState) {
            super.onCreate(savedInstanceState);
            setContentView(R.layout.activity_main);
            btnOk = findViewById(R.id.btn_ok);
            btnCancel = findViewById(R.id.btn_cancel);
            btnOk.setOnClickListener(this);
            btnCancel.setOnClickListener(this);
        }
        @Override
        public void onClick(View v) {
            switch (v.getId()){
                case R.id.btn_ok:
                    Toast.makeText(MainActivity.this,"您单击了确定按钮",Toast.LENGTH_SHORT).show();
                    break;
                case R.id.btn_cancel:
                    Toast.makeText(MainActivity.this,"您单击了取消按钮",Toast.LENGTH_SHORT).show();
                    break;
            }
        }
}
```

Button 控件的常用属性和方法如表 8.3 所示。

表 8.3 Button 控件的常用属性和方法

属　性	方　法	说　明
android:background	setBackgroundColor()	设置 Button 控件的背景颜色
android:text	setText()	设置 Button 控件的文本
android:textColor	setTextColor()	设置 Button 控件的文字颜色
android:textStyle		设置 Button 控件的文字格式，italic 表示斜体，bold 表示粗体
android:background	setBackgroundResource()	设置 Button 控件的背景图片

Button 控件的使用实例如图 8.5 所示。

图 8.5　Button 控件的使用实例

8.3.4　ImageButton 控件

ImageButton 控件继承自 ImageView，该控件既可以显示图片，又可以作为 Button 控件使用。与 Button 控件的区别是，ImageButton 控件没有 text 属性。

在 ImageButton 控件中显示的图片既可以通过 android:src 属性来设置，也可以通过 setImageResource(int)来设置。在默认情况下，ImageButton 控件与 Button 控件具有相同的背景颜色，当相应的按钮处于不同的状态时，背景颜色会发生变化。通常，将 ImageButton 控件的背景颜色设置为图片或者透明，以避免在 ImageButton 控件中显示的图片不能完全覆盖背景色，影响显示效果。

下面举例说明使用 XML 文件声明 ImageButton 控件的方法，代码如下：

```
<?xml version="1.0" encoding="utf-8"?>
<LinearLayout xmlns:android="http://schemas.android.com/apk/res/android"
    xmlns:tools="http://schemas.android.com/tools"
    android:layout_width="match_parent"
    android:layout_height="match_parent"
    android:orientation="vertical">
    <ImageButton
        android:layout_width="200dp"
        android:layout_height="200dp"
        android:background="@drawable/choujiang"            //choujiang 是图片“立即抽奖”的文件名
```

```
            android:scaleType="centerInside"
        />
</LinearLayout>
```

ImageButton 控件的使用实例如图 8.6 所示。

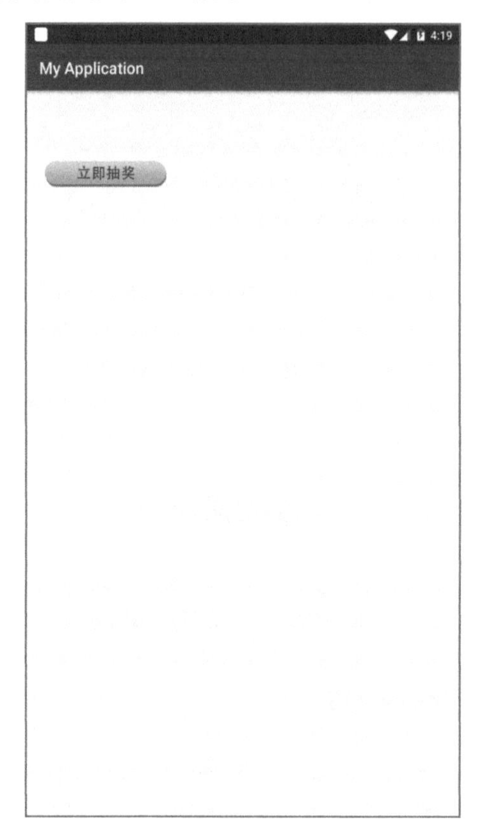

图 8.6　ImageButton 控件的使用实例

8.3.5　CheckBox 控件

CheckBox 控件是一个可以同时选择多个选项的控件（复选框），该控件继承自 android.widget.CompoundButton。CheckBox 控件的继承关系结构如图 8.7 所示。

```
Java.lang.Object
    →android.view.View
        →android.widget.TextView
            →android.widget.Button
                →android.widget.CompoundButton
                    →android.widget.CheckBox
```

图 8.7　CheckBox 控件的继承关系结构

CheckBox 控件的常用方法如表 8.4 所示。

表 8.4　CheckBox 控件的常用方法

方　法	说　明
isChecked()	检查选项是否被选中
setChecked(boolean)	如为 true，则设置成选中状态
setOnCheckedChangeListener()	处理 CheckBox 控件被选中事件，监听选项状态是否更改，把 CompoundButton.OnCheckedChangeListener 实例作为参数传入
getText()	获取复选框（CheckBox 控件）的值

（1）在 XML 文件中声明 CheckBox 控件的代码如下：

```
<?xml version="1.0" encoding="utf-8"?>
<LinearLayout xmlns:android="http://schemas.android.com/apk/res/android"
    xmlns:tools="http://schemas.android.com/tools"
    android:layout_width="match_parent"
    android:layout_height="match_parent"
    android:orientation="vertical"
    android:padding="20dp">
    <TextView
        android:layout_width="match_parent"
        android:layout_height="wrap_content"
        android:text="请选择您喜欢的运动:"
        android:textSize="20sp" />
    <CheckBox
        android:id="@+id/ckb_football"
        android:layout_width="wrap_content"
        android:layout_height="wrap_content"
        android:text="足球"
        android:textSize="20sp" />
    <CheckBox
        android:id="@+id/ckb_basketball"
        android:layout_width="wrap_content"
        android:layout_height="wrap_content"
        android:text="篮球"
        android:textSize="20sp" />
    <CheckBox
        android:id="@+id/ckb_running"
        android:layout_width="wrap_content"
        android:layout_height="wrap_content"
        android:text="跑步"
        android:textSize="20sp" />
    <CheckBox
        android:id="@+id/ckb_swiming"
        android:layout_width="wrap_content"
        android:layout_height="wrap_content"
        android:text="游泳"
        android:textSize="20sp" />
```

```
    <Button
        android:id="@+id/btn_ok"
        android:layout_width="wrap_content"
        android:layout_height="wrap_content"
        android:text="确定"
        android:textSize="20sp" />
</LinearLayout>
```

（2）引用 XML 文件中声明的 CheckBox 控件的代码如下：

```
public class MainActivity extends AppCompatActivity {
    private CheckBox ckbFootball;
    private CheckBox ckbBasketball;
    private CheckBox ckbSwimming;
    private CheckBox ckbRunning;
    private Button btnOk;
    @Override
    protected void onCreate(Bundle savedInstanceState) {
        super.onCreate(savedInstanceState);
        setContentView(R.layout.activity_main);
        ckbFootball = findViewById(R.id.ckb_football);
        ckbBasketball = findViewById(R.id.ckb_basketball);
        ckbSwimming = findViewById(R.id.ckb_swiming);
        ckbRunning = findViewById(R.id.ckb_running);
        btnOk = findViewById(R.id.btn_ok);
        btnOk.setOnClickListener(new View.OnClickListener() {
            @Override
            public void onClick(View v) {
                StringBuffer buffer = new StringBuffer();
                buffer.append("您喜欢的运动有:");
                if (ckbFootball.isChecked()) {
                    buffer.append("足球、");
                }
                if (ckbBasketball.isChecked()) {
                    buffer.append("篮球、");
                }
                if (ckbSwimming.isChecked()) {
                    buffer.append("游泳、");
                }
                if (ckbRunning.isChecked()) {
                    buffer.append("跑步、");
                }
                String result = buffer.substring(0, buffer.length() - 1).concat("。");
                Toast.makeText(MainActivity.this, result, Toast.LENGTH_SHORT).show();
            }
        });
    }
}
```

CheckBox 控件的使用实例如图 8.8 所示。

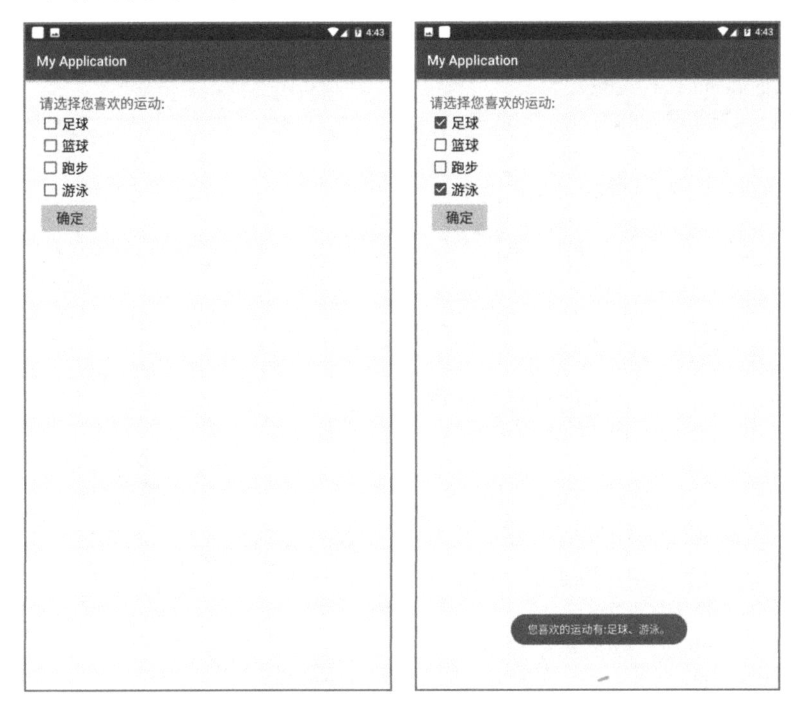

图 8.8　CheckBox 控件的使用实例

8.3.6　RadioButton 控件

RadioButton 是一个用于选择一个选项的控件（单选框），该控件继承自 android.widget.CompoundButton。RadioGroup 是 RadioButton 的承载体，在程序运行时不可见，在应用程序中可能包含一个或多个 RadioGroup。RadioGroup 是线性布局（LinearLayout）的子类，一个 RadioGroup 可以包含多个 RadioButton 控件。RadioGroup 用于对单选框进行分组，在每个 RadioGroup 中（相同组内的单选框），用户仅能够选择其中一个 RadioButton 控件。RadioButton 控件的继承关系结构如图 8.9 所示。

Java.lang.Object
　　→android.view.View
　　　　→android.widget.ViewGroup
　　　　　　→android.widget.LinearLayout
　　　　　　　　→android.widget.RadioGroup

图 8.9　RadioButton 控件的继承关系结构

如果要监听单选框状态的更改，则需要在 RadioGroup 中添加 setOnCheckedChangeListener(RadioGroup.OnCheckedChangeListener)监听器。

（1）在 XML 文件中声明 RadioGroup 和 RadioButton 控件的代码如下：

```
<?xml version="1.0" encoding="utf-8"?>
<LinearLayout xmlns:android="http://schemas.android.com/apk/res/android"
```

```
    xmlns:tools="http://schemas.android.com/tools"
    android:layout_width="match_parent"
    android:layout_height="match_parent"
    android:orientation="vertical"
    android:padding="20dp">
    <TextView
        android:layout_width="match_parent"
        android:layout_height="wrap_content"
        android:text="请选择您的性别:"
        android:textSize="20sp" />
    <RadioGroup
        android:layout_width="wrap_content"
        android:layout_height="wrap_content"
        android:orientation="horizontal">
        <RadioButton
            android:id="@+id/rbt_boy"
            android:layout_width="wrap_content"
            android:layout_height="wrap_content"
            android:text="男"
            android:textSize="20sp" />
        <RadioButton
            android:layout_width="wrap_content"
            android:layout_height="wrap_content"
            android:text="女"
            android:textSize="20sp" />
    </RadioGroup>
    <Button
        android:id="@+id/btn_ok"
        android:layout_width="wrap_content"
        android:layout_height="wrap_content"
        android:text="确定"
        android:textSize="20sp" />
</LinearLayout>
```

（2）引用 RadioGroup 和 RadioButton 控件的代码如下：

```
public class MainActivity extends AppCompatActivity {
    private Button btnOk;
    private RadioButton rbtBoy;
    @Override
    protected void onCreate(Bundle savedInstanceState) {
        super.onCreate(savedInstanceState);
        setContentView(R.layout.activity_main);
        rbtBoy = findViewById(R.id.rbt_boy);
        btnOk = findViewById(R.id.btn_ok);
        btnOk.setOnClickListener(new View.OnClickListener() {
            @Override
            public void onClick(View v) {
```

```
                if (rbtBoy.isChecked()) {
                    Toast.makeText(MainActivity.this, "您的性别是男", Toast.LENGTH_SHORT).show();
                } else {
                    Toast.makeText(MainActivity.this, "您的性别是女", Toast.LENGTH_SHORT).show();
                }
            }
        });
    }
}
```

RadioButton 控件和 RadioGroup 的常用方法如表 8.5 所示。

表 8.5 RadioButton 控件和 RadioGroup 的常用方法

方　法	说　明
RadioGroup.check(int id)	通过传递的参数来设置 RadioButton 控件
RadioGroup.clearCheck()	清空选中的选项
RadioGroup.setCheckedChangeListener()	处理选项被选中事件，把 RadioGroup.OnCheckedChangeListener 实例作为参数传入
RadioButton.getText()	获取单选框（RadioButton 控件）的值

RadioButton 控件的使用实例如图 8.10 所示。

图 8.10 RadioButton 控件的使用实例

8.3.7 Spinner 控件

Spinner 控件是一种下拉列表，其数据通过 ArrayAdapter 提供。ArrayAdapter 提供数据的方式有两种，一种是在资源文件中提供数据，另一种是在程序中构造数组。Spinner 控件的部分属性和方法如表 8.6 所示。

表 8.6　Spinner 控件的部分属性和方法

属　性	方　法	说　明
android:dropDownHorizontalOffset	setDropDownHorizontalOffset(int pixels)	当 Spinner 控件的模式为 DROPDOWN 时，设置下拉列表的水平偏移
android:dropDownSelector		当 Spinner 控件的模式为 DROPDOWN 时，设置下拉列表的显示效果
android:dropDownVerticalOffset	setDropDownVerticalOffset(int pixels)	当 Spinner 控件的模式为 DROPDOWN 时，设置下拉列表的竖直偏移
android:dropDownWidth	setDropDownWidth(int)	当 Spinner 控件的模式为 DROPDOWN 时，设置下拉列表的宽度
android:gravity	setGravity(int)	设置当前选中项的对齐方式

在 XML 文件中声明 Spinner 控件的代码如下：

```
<?xml version="1.0" encoding="utf-8"?>
<LinearLayout xmlns:android="http://schemas.android.com/apk/res/android"
    xmlns:app="http://schemas.android.com/apk/res-auto"
    xmlns:tools="http://schemas.android.com/tools"
    android:layout_width="match_parent"
    android:layout_height="match_parent"
    android:orientation="vertical"
    tools:context=".MainActivity">
    <TextView
        android:layout_width="match_parent"
        android:layout_height="wrap_content"
        android:text="请选择您的专业"
        android:textSize="20sp" />
    <Spinner
        android:id="@+id/sp_major"
        android:layout_width="wrap_content"
        android:layout_height="wrap_content"
        android:entries="@array/major" />
</LinearLayout>
```

Spinner 控件中使用的数据需要存储在一个数组中。在“res/values”目录下新建一个 arrays.xml 文件，代码如下：

```
<?xml version="1.0" encoding="utf-8"?>
<resources>
    <string-array name="major">
        <item>软件</item>
        <item>网络</item>
        <item>大数据</item>
        <item>云计算</item>
        <item>人工智能</item>
```

```
    </string-array>
</resources>
```

引用 Spinner 控件的代码如下：

```
public class SpinnerActivity extends AppCompatActivity {
    private Spinner spinner;
    @Override
    protected void onCreate(Bundle savedInstanceState) {
        super.onCreate(savedInstanceState);
        setContentView(R.layout.activity_spinner);
        spinner = findViewById(R.id.sp_major);
        spinner.setOnItemSelectedListener(new AdapterView.OnItemSelectedListener() {
            @Override
            public void onItemSelected(AdapterView<?> parent, View view, int position, long id) {
                String major = getResources().getStringArray(R.array.major)[position];
                Toast.makeText(SpinnerActivity.this,
                        "您的专业是" + major, Toast.LENGTH_SHORT).show();
            }
            @Override
            public void onNothingSelected(AdapterView<?> parent) {
            }
        });
    }
}
```

Spinner 控件的使用实例如图 8.11 所示。

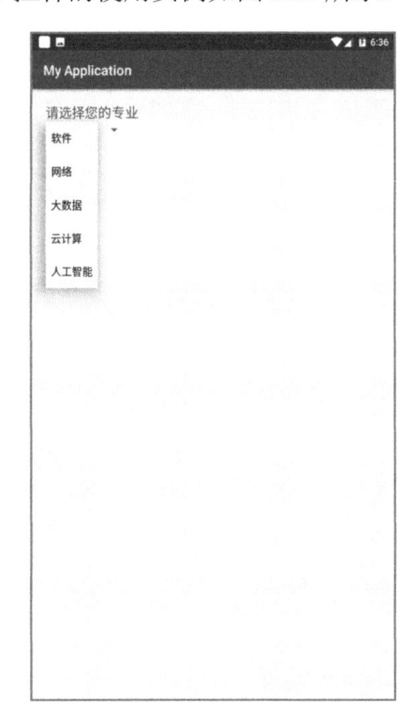

图 8.11　Spinner 控件的使用实例

8.3.8 ListView 控件

ListView 是一种常用于竖直显示的列表控件，能够以列表的形式显示具体内容。如果 ListView 控件显示内容过多，则会出现竖直滚动条，并且该滚动条能够自适应数据的长度。ListView 控件如图 8.12 所示。

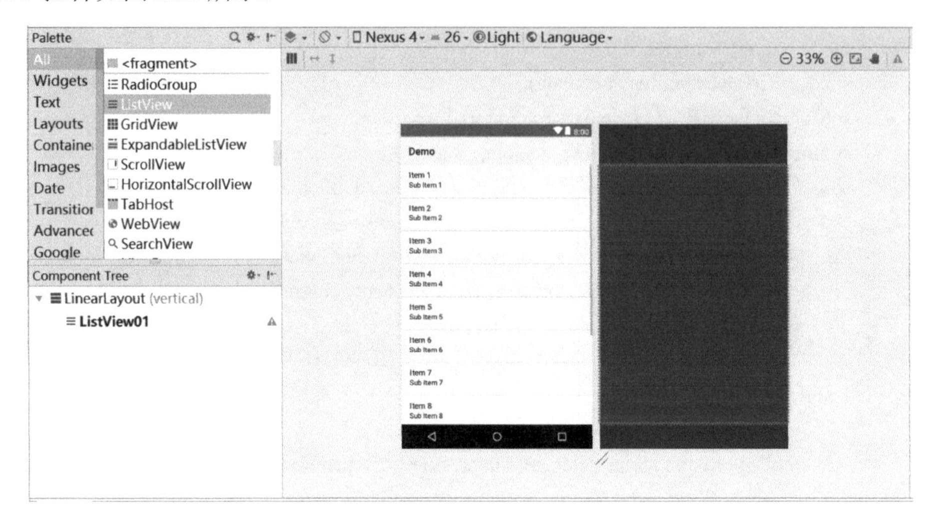

图 8.12　ListView 控件

根据适配器的类型，ListView 控件可以分为 ArrayAdapter、SimpleAdapter 和 SimpleCursorAdapter 三种。ArrayAdapter 只能展示一行文本；SimpleAdapter 有较好的扩展性，可以自定义各种效果；SimpleCursorAdapter 是 SimpleAdapter 对数据库的简单结合，可以将数据库中的内容以列表的形式显示出来。

ListView 控件支持单击事件处理，实现较强大的选择功能。ListView 控件的常用属性如表 8.7 所示。

表 8.7　ListView 控件的常用属性

属　性	说　明
android:dividerHeight	分隔符的高度。若没有指明高度，则用此分隔符固有的高度。高度必须是带单位的浮点数，如 14.5 sp，可用的单位有 px、dp、sp、in、mm 等
android:entries	引用一个将在 ListView 控件中使用的数组。若数组是固定的，使用该属性比在程序中定义数组更为简单
android:footerDividersEnabled	当设成 false 时，此 ListView 控件将不会在页脚视图前画分隔符。该属性的默认值为 true，属性值必须设置为 true 或 false
android:headerDividersEnabled	当设成 false 时，此 ListView 控件将不会在页眉视图后画分隔符。该属性的默认值为 true，属性值必须设置为 true 或 false
android:choiceMode	规定 ListView 控件所使用的选择模式。在默认状态下，没有选择模式。该属性必须设置为以下常量之一：none，值为 0，表示无选择模式；singleChoice，值为 1，表示最多可以选中一项；multipleChoice，值为 2，表示可以选中多项

在 XML 文件中声明 ListView 控件的代码如下：

```
<?xml version="1.0" encoding="utf-8"?>
<LinearLayout xmlns:android="http://schemas.android.com/apk/res/android"
    xmlns:app="http://schemas.android.com/apk/res-auto"
    xmlns:tools="http://schemas.android.com/tools"
    android:layout_width="match_parent"
    android:layout_height="match_parent"
    tools:context=".ListViewActivity">
    <ListView
        android:layout_width="match_parent"
        android:layout_height="match_parent"
        android:entries="@array/major" />
</LinearLayout>
```

ListView 控件的使用示例如图 8.13 所示。

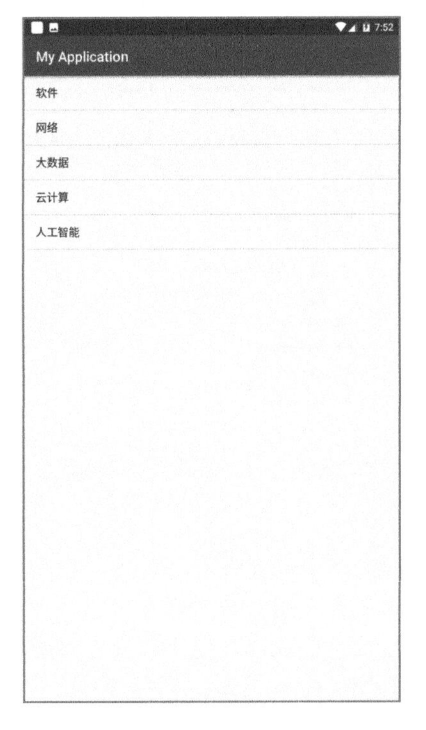

图 8.13　ListView 控件的使用示例

8.4　开发实践：城市气象监控设备管理系统应用界面的设计

8.4.1　开发设计

本任务设计一个城市气象监控设备管理系统的应用界面，包括设备的显示、增加和删除等功能，使用 ListView 控件显示所有的监控设备，使用 Button 控件触发监控设备的添加功能。城市气象监控设备管理系统的应用界面如图 8.14 所示。

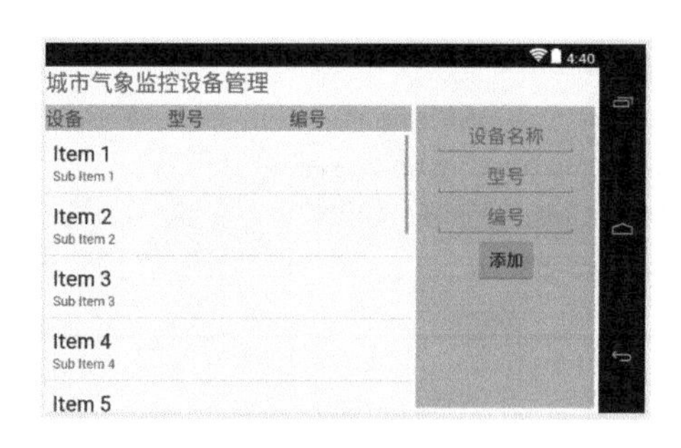

图 8.14　城市气象监控设备管理系统的应用界面

本项目主要编写 MainActivity.java、activity_main.xm 和 list_person_layout.xml 等文件。城市气象监控设备管理系统应用界面的目录结构如图 8.15 所示。

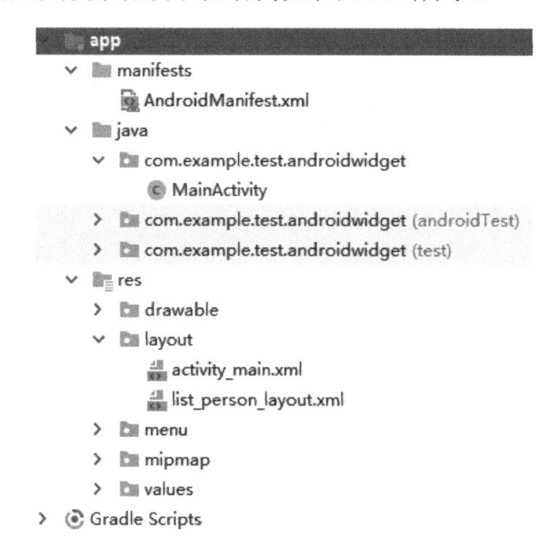

图 8.15　城市气象监控设备管理系统应用界面的目录结构

项目开发步骤如下：

（1）编写 list_person_layout.xml 布局文件。

（2）编写 activity_main.xml 布局文件，设置布局管理器，添加应用界面上使用的控件。

（3）编写 MainActivity.java 文件的代码，定义界面控件对象。

（4）在 onCreate 方法中通过 findViewById 实例化控件。

（5）通过设置 setOnItemLongClickListener 监听器，实现长按删除 ListView 控件中的条目。

（6）定义动态数组 list 来存储数据，通过适配器 SimpleAdapter 来绑定数据。

（7）通过实例化 insertBtn 来添加按钮，设置按钮的 setOnClickListener 监听器，响应用户单击按钮事件。

（8）在 onClick 方法中判断用户的数据输入，将有效的数据添加到 ListView 控件。

8.4.2　功能实现

1．布局设计

本项目的布局用到了 Android 的 ListView、TextView、Button 等控件，具体代码在

activity_main.xml 布局文件中，如下所示：

```
<RelativeLayout xmlns:android="http://schemas.android.com/apk/res/android"
    xmlns:tools="http://schemas.android.com/tools"
    android:id="@+id/container"
    android:layout_width="match_parent"
    android:layout_height="match_parent"
    tools:ignore="MergeRootFrame">
    <!-- 顶部  -->
    <RelativeLayout
        android:id="@+id/MyLayout_top"
        android:layout_width="fill_parent"
        android:layout_height="40dp"
        android:layout_alignParentTop="true"
        android:orientation="horizontal">
        <!-- 标题 -->
        <TextView
            android:layout_width="match_parent"
            android:layout_height="wrap_content"
            android:layout_weight="1"
            android:text="城市气象监控设备管理"
            android:textSize="24sp" />
    </RelativeLayout>
    <!-- 显示列表 -->
    <RelativeLayout
        android:id="@+id/MyLayout_left"
        android:layout_width="match_parent"
        android:layout_height="match_parent"
        android:layout_marginRight="200dp"
        android:layout_below="@id/MyLayout_top"
        android:orientation="horizontal">
        <LinearLayout
            android:id="@+id/list_header"
            android:layout_width="match_parent"
            android:layout_height="wrap_content"
            android:gravity="center"
            android:background="@color/gray"
            android:orientation="horizontal">
            <TextView
                android:layout_width="match_parent"
                android:layout_height="wrap_content"
                android:layout_weight="1"
                android:text="设备"
                android:textSize="20sp" />
            <TextView
                android:layout_width="match_parent"
                android:layout_height="wrap_content"
```

```
                android:layout_weight="1"
                android:text="型号"
                android:textSize="20sp" />
            <TextView
                android:layout_width="match_parent"
                android:layout_height="wrap_content"
                android:layout_weight="1"
                android:text="编号"
                android:textSize="20sp" />
        </LinearLayout>
        <!-- 显示表内容 -->
        <ListView
            android:id="@+id/listview1"
            android:layout_below="@+id/list_header"
            android:layout_width="match_parent"
            android:layout_height="wrap_content"
            android:gravity="center"></ListView>
    </RelativeLayout>
    <RelativeLayout
        android:id="@+id/MyLayout_right"
        android:layout_width="190dp"
        android:layout_marginLeft="5dp"
        android:layout_marginRight="5dp"
        android:layout_marginBottom="5dp"
        android:layout_height="match_parent"
        android:layout_alignParentRight="true"
        android:layout_below="@id/MyLayout_top"
        android:gravity="center">
        <LinearLayout
            android:layout_width="fill_parent"
            android:layout_height="fill_parent"
            android:background="@color/gray"
            android:layout_alignParentBottom="true"
            android:orientation="vertical">
            <LinearLayout
                android:layout_width="fill_parent"
                android:layout_height="wrap_content"
                android:layout_marginTop="10dp"
                android:gravity="center"
                android:orientation="vertical">
                <EditText
                    android:id="@+id/edit_name"
                    android:layout_width="150dp"
                    android:layout_height="wrap_content"
                    android:layout_weight="1"
                    android:gravity="center"
                    android:hint="设备名称"
```

```
                    android:textSize="20sp" />
                <EditText
                    android:id="@+id/edit_gender"
                    android:layout_width="150dp"
                    android:layout_height="wrap_content"
                    android:layout_weight="1"
                    android:gravity="center"
                    android:hint="型号"
                    android:textSize="20sp" />
                <EditText
                    android:id="@+id/edit_cardNo"
                    android:layout_width="150dp"
                    android:layout_height="wrap_content"
                    android:layout_weight="1"
                    android:gravity="center"
                    android:hint="编号"
                    android:textSize="20sp" />
                <Button
                    android:id="@+id/insert_btn"
                    android:layout_width="wrap_content"
                    android:layout_height="match_parent"
                    android:layout_weight="1"
                    android:text="添加" />
            </LinearLayout>
        </LinearLayout>
    </RelativeLayout>
</RelativeLayout>
```

2．列表布局文件设计

```
<?xml version="1.0" encoding="utf-8"?>
<LinearLayout xmlns:android="http://schemas.android.com/apk/res/android"
    android:layout_width="match_parent"
    android:layout_height="match_parent"
    android:orientation="horizontal" >
    <TextView
        android:id="@+id/name"
        android:layout_width="match_parent"
        android:layout_height="wrap_content"
        android:layout_weight="1"
        android:textSize="20sp" />
    <TextView
        android:id="@+id/gender"
        android:layout_width="match_parent"
        android:layout_height="wrap_content"
        android:layout_weight="1"
        android:textSize="20sp" />
```

```
        <TextView
            android:id="@+id/cardNo"
            android:layout_width="match_parent"
            android:layout_height="wrap_content"
            android:layout_weight="1"
            android:textSize="20sp" />
</LinearLayout>
```

3. 主程序设计

```
public class MainActivity extends Activity {
    List<Map<String, Object>> list;
    SimpleAdapter adapter;
    private ListView listview;
    private ContentValues values;
    private EditText nameET;
    private EditText genderET;
    private EditText cardNoET;
    private Button insertBtn;
    @Override
    protected void onCreate(Bundle savedInstanceState) {
        super.onCreate(savedInstanceState);
        //去掉 TitleBar
        requestWindowFeature(Window.FEATURE_NO_TITLE);
        setContentView(R.layout.activity_sensor_guard);
        nameET = (EditText) findViewById(R.id.edit_name);
        genderET = (EditText) findViewById(R.id.edit_gender);
        cardNoET = (EditText) findViewById(R.id.edit_cardNo);
        listview = (ListView) findViewById(R.id.listview1);
        listview.setOnItemLongClickListener(new AdapterView.OnItemLongClickListener() {
            @Override
            public boolean onItemLongClick(AdapterView<?> arg0, View arg1,
                                            int arg2, long arg3) {
                list.remove(arg2);
                adapter.notifyDataSetChanged();
                Toast.makeText(MainActivity.this, "第" + (arg2 + 1) + "行已删除", Toast.LENGTH_
SHORT).show();
                return true;
            }
        });
        //定义动态数组 list 存储数据
        list = new ArrayList<Map<String, Object>>();
        //通过适配器 SimpleAdapter 绑定数据
        //构造函数 SimpleAdapter 未定义，需把要 this 修改为 MainActivity.this
        adapter = new SimpleAdapter(this, list, R.layout.list_person_layout,
                new String[]{"name", "gender", "cardNo"},
                new int[]{R.id.name, R.id.gender, R.id.cardNo});
```

```
            listview.setAdapter(adapter);
            insertBtn = (Button) findViewById(R.id.insert_btn);
            insertBtn.setOnClickListener(new OnClickListener() {
                @Override
                public void onClick(View v) {
                    String name = nameET.getText().toString().trim();
                    String gender = genderET.getText().toString().trim();
                    String no = cardNoET.getText().toString().trim();
                    if (name.length() == 0 || no.length() == 0) {
                        Toast.makeText(MainActivity.this, "气象监控设备名称和编号不能为空！", Toast.
LENGTH_SHORT).show();
                        return;
                    }
                    Map<String, Object> map = new HashMap<String, Object>();
                    map.put("name", name);
                    map.put("gender", gender);
                    map.put("cardNo", no);
                    list.add(map);
                    adapter.notifyDataSetChanged();
                    nameET.setText("");
                    genderET.setText("");
                    cardNoET.setText("");
                }
            });
        }
        protected void onDestroy() {
            super.onDestroy();}
    }
```

8.5　任务验证

在 Android Studio 开发环境中打开本项目的例程，编译通过后运行程序。程序运行效果如图 8.16 所示，通过右侧编辑栏可输入监控设备信息。

图 8.16　程序运行效果

单击“添加”按钮可以添加监控设备的信息，监控设备信息保存在动态数组 list 中，并在 ListView 控件中显示，如图 8.17 所示。长按 ListView 控件中的条目可以删除相应的监控设备，如图 8.18 所示。

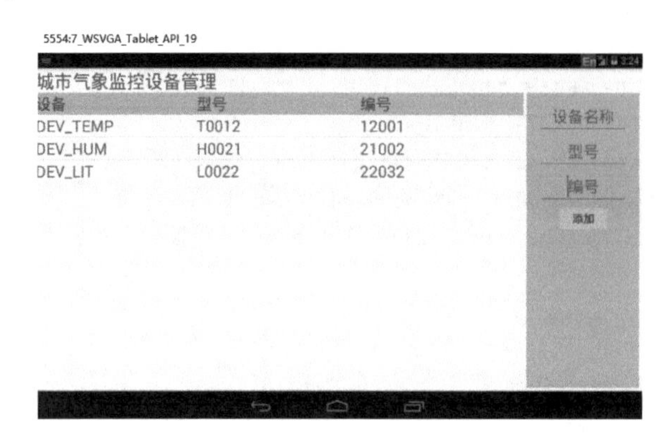

图 8.17　ListView 控件中显示的监控设备信息

图 8.18　长按 ListView 控件中的条目可删除相应的监控设备

8.6　开发小结

本任务主要介绍 Android 界面控件的基础知识。通过本任务的学习，读者可以掌握 Android 界面控件的使用方法，完成城市气象监控设备管理系统应用界面的设计。

8.7　思考与拓展

（1）TextView 控件和 EditText 控件有哪些区别？

（2）请尝试给本项目中的 ListView 控件增加红色的显示背景。

任务9 城市环境系统功能菜单的设计

本任务通过学习 Android 选项菜单、子菜单、上下文菜单，帮助读者掌握 Android 菜单的使用方法，从而实现城市环境系统功能菜单的设计。

9.1 开发场景：如何为城市环境系统增加功能菜单

近些年，随着互联网的普遍应用，以及城市居民生活水平的提升，政府对城市管理的智能化，尤其是智能城市环境监测管理日益重视。本任务假设完成了城市环境系统的开发，现需要增加两个功能菜单，一个是“城市空气”，另一个是“城市温度”，这两个功能菜单又分别包含子菜单。该如何实现呢？

9.2 开发目标

（1）知识目标：熟悉 Android 菜单。
（2）技能目标：掌握 Android 菜单的使用方法。
（3）任务目标：使用 Android 菜单，实现城市环境系统功能菜单的设计。

9.3 原理学习：熟悉 Android 菜单

在不占用界面空间的前提下，Android 菜单可以为应用程序提供统一的功能和设置界面，并为开发人员提供易于使用的编程接口。Android 菜单不仅可以在应用程序中定义，还可以像界面布局一样在 XML 文件中定义。使用 XML 文件定义 Android 菜单，可以将代码与界面进行分类设计，简化代码的复杂度，有利于界面的可视化。

Android 菜单包括三类：选项菜单、子菜单和上下文菜单。

在 Activity 中可以通过覆写 onCreateOptionsMenu(Menu menu)方法创建选项菜单。在用户按下手机界面中的 Menu 按钮时会显示创建好的菜单。在 onCreateOptionsMenu(Menu menu)方法的内部可以调用 Menu.add()方法来实现菜单的添加。

如果要处理选中事件，则可以通过覆写 Activity 中的 onMenuItemSelected()方法来实现，该方法常用于处理菜单选中事件。

9.3.1 Android 的选项菜单

选项菜单可分为两类：图标菜单（Icon Menu）和扩展菜单（Expanded Menu）。

在 Android 4.0 以后的版本中，图标菜单默认为竖直的列表菜单，可以同时显示文字和图标。图标菜单不支持单选框和复选框。在创建选项菜单时，如果没有采用在上述 XML 文件中设定显示原有风格的方法，而仅通过 setIcon()方法给选项菜单添加图标，则图标是无法显示出来的。其原因是在 Android 4.0 以后的版本中，涉及选项菜单的源码类 MenuBuilder 做了改变，mOptionalIconsVisible 初始默认为 false。只要在创建选项菜单时通过调用 setOptionalIconsVisible()方法将 mOptionalIconsVisible 设置为 true 即可显示。

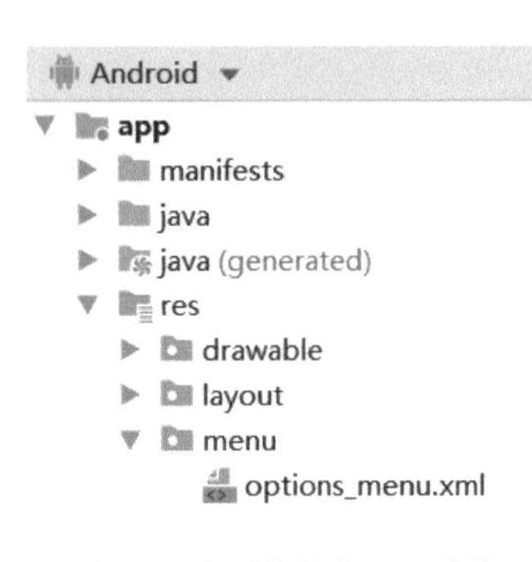

图 9.1 选项菜单的目录结构

扩展菜单是竖直的列表菜单，它不支持显示图标，但支持单选框和复选框。

只有在 Activity 中重载 onCreateOptionMenu()方法，才能在 Android 应用程序中使用选项菜单。在第一次使用选项菜单时，首先要在 res 中新建 menu 目录，然后创建 options_menu.xml 文件。选项菜单的目录结构如图 9.1 所示。

```
<menu xmlns:android="http://schemas.android.com/apk/res/android"
    xmlns:app="http://schemas.android.com/apk/res-auto"
    xmlns:tools="http://schemas.android.com/tools"
    tools:context="com.android.peter.menudemo.MenuDemoActivity">
    <item
        android:id="@+id/action_1"
        android:orderInCategory="100"
        android:title="选项菜单 1"
        app:showAsAction="never" />
    <item
        android:id="@+id/action_2"
        android:orderInCategory="100"
        android:title="选项菜单 2"
        app:showAsAction="never" />
    <item
        android:id="@+id/action_3"
        android:orderInCategory="100"
        android:title="选项菜单 3"
        app:showAsAction="never" />
</menu>
```

在 Activity 中重载 onCreateOptionMenu()方法，加载上面创建的 options_menu.xml 文件，通过重载 onOptionsItemSelected()方法对选项菜单的选中事件进行处理。

```
public class MenuActivity extends AppCompatActivity {
    @Override
    protected void onCreate(Bundle savedInstanceState) {
        super.onCreate(savedInstanceState);
        setContentView(R.layout.activity_menu);
    }
    //创建选项菜单，加载创建的 options_menu.xml 布局文件
    @Override
    public boolean onCreateOptionsMenu(Menu menu) {
        getMenuInflater().inflate(R.menu.options_menu, menu);
        return true;
    }
    //当 OptionsMenu 被选中时处理具体的响应事件
    @Override
    public boolean onOptionsItemSelected(MenuItem item) {
        switch (item.getItemId()) {
            case R.id.action_1:
                Toast.makeText(MenuActivity.this, "选项菜单 1", Toast.LENGTH_SHORT).show();
                return true;
            case R.id.action_2:
                Toast.makeText(MenuActivity.this, "选项菜单 2", Toast.LENGTH_SHORT).show();
                return true;
            case R.id.action_3:
                Toast.makeText(MenuActivity.this, "选项菜单 3", Toast.LENGTH_SHORT).show();
                return true;
        }
        return super.onOptionsItemSelected(item);
    }
}
```

选项菜单的使用实例如图 9.2 所示。

图 9.2　选项菜单的使用实例

9.3.2 Android 的子菜单

在子菜单（SubMenu）中，菜单选项使用浮动窗体的形式显示，可以更好地适应小屏幕的显示。Android 系统可以在选项菜单或快捷菜单中使用子菜单，有利于将相同或相似的菜单选项组织在一起，便于显示和分类。子菜单是通过 addSubMenu()方法来添加的。

```
SubMenu uploadMenu = (SubMenu) menu.addSubMenu(0,UPLOAD,1, "城市空气").setIcon
(R.drawable.upload);
uploadMenu.setHeaderIcon(R.drawable.upload);
uploadMenu.setHeaderTitle("空气系统设置");
uploadMenu.add(0,SUB_UPLOAD_A,0, "设置参数 A");
uploadMenu.add(0,SUB_UPLOAD_B,0, "设置参数 B");
```

第 1 行代码在 onCreateOptionsMenu()方法传递的 menu 对象中调用 addSubMenu()方法，在选项菜单中添加一个子菜单，用户单击选项菜单后可以打开子菜单。

第 2 行代码调用 setHeaderIcon()方法来定义子菜单的图标。

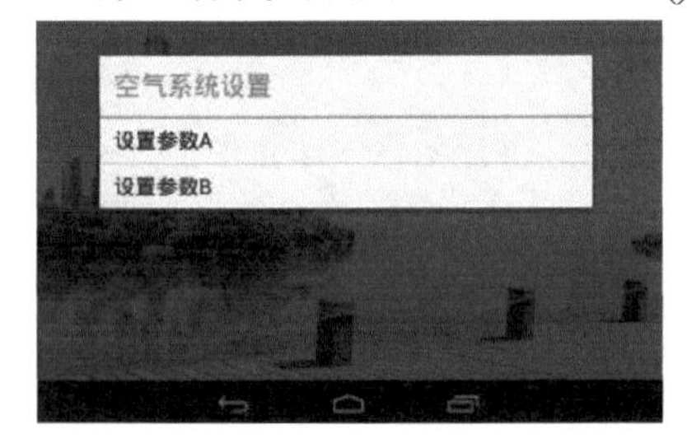

图 9.3　子菜单的使用实例

第 3 行代码定义子菜单的标题，如果不设定子菜单的标题，子菜单将显示父菜单选项标题，即第 1 行代码中的“城市空气”。

第 4 行代码和第 5 行代码在子菜单中添加了两个菜单选项，菜单选项的更新和选中事件处理仍然使用 onPrepareOptionsMenu()方法和 onOptionsItemSelected()方法。

子菜单的使用实例如图 9.3 所示。

9.3.3 Android 的上下文菜单

上下文菜单（ContextMenu）又称为内容菜单，当长按 ContextMenu 控件时，会弹出一个上下文菜单。上下文菜单是由选项菜单扩展而来的。

在 TextView 控件上添加上下文菜单的方法为：首先在布局的 XML 文件中增加一个 TextView 控件并实例化，然后为 TextView 控件注册上下文菜单，最后通过 onCreateContextMenu()方法创建上下文菜单。当长按 TextView 控件时，onCreateContextMenu()方法会把定义好的 XML 文件填充进来。菜单资源文件 context_menu.xml 内容如下：

```
<menu xmlns:android="http://schemas.android.com/apk/res/android"
    xmlns:app="http://schemas.android.com/apk/res-auto"
    xmlns:tools="http://schemas.android.com/tools"
    tools:context="com.android.peter.menudemo.MenuDemoActivity">
    <item
        android:id="@+id/action_1"
        android:orderInCategory="100"
        android:title="上下文菜单 1"
        app:showAsAction="never" />
```

```
        <item
            android:id="@+id/action_2"
            android:orderInCategory="100"
            android:title="上下文菜单 1"
            app:showAsAction="never" />
        <item
            android:id="@+id/action_3"
            android:orderInCategory="100"
            android:title="上下文菜单 3"
            app:showAsAction="never" />
    </menu>
```

TextView 控件的布局文件如下：

```
<?xml version="1.0" encoding="utf-8"?>
<LinearLayout xmlns:android="http://schemas.android.com/apk/res/android"
    xmlns:app="http://schemas.android.com/apk/res-auto"
    xmlns:tools="http://schemas.android.com/tools"
    android:layout_width="match_parent"
    android:layout_height="match_parent"
    android:padding="30dp"
    tools:context=".ContextMenuActivity">
    <TextView
        android:id="@+id/textview"
        android:layout_width="wrap_content"
        android:layout_height="wrap_content"
        android:layout_centerInParent="true"
        android:longClickable="true"
        android:text="长按弹出上下文菜单"
        android:textSize="20sp" />
</LinearLayout>
```

Activity 中 TextView 控件初始化和上下文菜单创建的代码如下：

```
public class ContextMenuActivity extends AppCompatActivity {
    private TextView textView;
    @Override
    protected void onCreate(Bundle savedInstanceState) {
        super.onCreate(savedInstanceState);
        setContentView(R.layout.activity_context_menu);
        textView = findViewById(R.id.textview);
        registerForContextMenu(textView);
    }
    @Override
    public void onCreateContextMenu(ContextMenu menu, View v, ContextMenu.ContextMenuInfo
menuInfo) {
        super.onCreateContextMenu(menu, v, menuInfo);
        getMenuInflater().inflate(R.menu.context_menu, menu);
```

```
        }
    }
```

添加菜单选择监听事件，也就是覆写 onContextItemSelected()方法，当上下文菜单被选中时，该方法会被调用。代码如下：

```
@Override
    public boolean onContextItemSelected(MenuItem item) {
        switch (item.getItemId()) {
            case R.id.action_1:
                Toast.makeText(ContextMenuActivity.this, "选项菜单 1", Toast.LENGTH_SHORT).show();
                return true;
            case R.id.action_2:
                Toast.makeText(ContextMenuActivity.this, "选项菜单 2", Toast.LENGTH_SHORT).show();
                return true;
            case R.id.action_3:
                Toast.makeText(ContextMenuActivity.this, "选项菜单 3", Toast.LENGTH_SHORT).show();
                return true;
        }
        return super.onContextItemSelected(item);
    }
```

从上面可以看出，为 TextView 控件增加上下文菜单的过程可分为三个步骤。

第一步：覆写 onCreateContextMenu()方法，静态或者动态地创建上下文菜单。

第二步：覆写 onOptionsItemSelected()方法，为上下文菜单添加监听函数。

第三步：为 TextView 控件注册上下文菜单，这里的注册函数 registerForContextMenu()是将 TextView 控件和上下文菜单联系起来的桥梁。

上下文菜单的使用实例如图 9.4 所示。

图 9.4　上下文菜单的使用实例

9.4　开发实践：城市环境系统功能菜单

9.4.1　开发设计

本任务在已开发的城市环境系统中增加功能菜单，使用的是 Android 的选项菜单和子菜单。城市环境系统的两个主菜单采用 OptionMenu 控件，子菜单采用 SubMenu 控件。城市环境系统增加的主菜单如图 9.5 所示。

图 9.5　城市环境系统增加的主菜单

本任务的主要工作编写 MainActivity.java 文件，项目目录结构如图 9.6 所示。

图 9.6　项目目录结构

项目开发步骤如下：

（1）编写 activity_main.xml 文件，设置界面显示的标题。

（2）在 MainActivity.java 文件中定义表示菜单选项 ID 的变量。

（3）在 onCreate()方法中通过 setContentView 设置显示视图。

（4）覆写 onCreateOptionsMenu()方法，初始化选项菜单时会调用该方法。

（5）在 onCreateOptionsMenu()方法中，通过 menu.addSubMenu 添加子菜单 menu1 与 menu2，通过子菜单的 add()方法为菜单添加二级子菜单，此时 menu1 与 menu2 相当于父菜单。

（6）覆写 onOptionsItemSelected()方法，当菜单选项被选中后会调用该方法。

（7）在 onOptionsItemSelected()方法中通过 item.getItemId()该方法获取菜单选项 ID，进行选项计数处理。

9.4.2 功能实现

1. 布局设计

```
<RelativeLayout xmlns:android="http://schemas.android.com/apk/res/android"
    xmlns:tools="http://schemas.android.com/tools"
    android:layout_width="match_parent"
    android:layout_height="match_parent"
    android:background="@drawable/chshhjxt">
    <TextView
        android:layout_width="wrap_content"
        android:layout_height="wrap_content"
        android:layout_centerHorizontal="true"
        android:layout_centerVertical="true"
        android:text="城市环境系统功能菜单" />
</RelativeLayout>
```

2. 主程序设计

```
public class SubMenuActivity extends Activity {
    //用来表示菜单选项被选中的次数
    static int Menu1Counter = 0;
    static int Menu2Counter = 0;
    static final int MENU_01 = Menu.FIRST;
    static final int MENU_02 = Menu.FIRST + 1;
    //用来表示 MENU_UPLOAD 的两个子菜单选项 ID
    static final int SUB_MENU_UPLOAD_A = Menu.FIRST+2;
    static final int SUB_MENU_UPLOAD_B = Menu.FIRST+3;
    static final int SUB_MENU_UPLOAD_C = Menu.FIRST+4;
    static final int SUB_MENU_UPLOAD_D = Menu.FIRST+5;
    protected void onCreate(Bundle savedInstanceState) {
        super.onCreate(savedInstanceState);
        setContentView(R.layout.activity_sub_menu);
    }
    //使用选项菜单时，必须重载以下方法
    //在初始化选项菜单时，会调用 onCreateOptionsMenu()方法，该方法通常用来初始化菜单选项的相
    //关内容，设置菜单选项自身的子项 ID 和组 ID
    @Override
     public boolean onCreateOptionsMenu(Menu menu) {
        //第 1 个参数 groupId 是组 ID
        //第 2 个参数 itemid 是子项 ID，是每一个菜单选项的唯一标识
        //第 3 个参数 order 用于定义菜单选项在选项菜单中排列的先后顺序
        //第 4 个参数是菜单选项显示的标题
        //为菜单添加子菜单，此时 menu1 相当于父菜单
        SubMenu menu1 = (SubMenu) menu.addSubMenu(0, MENU_01, 1,"城市空气");
        menu1.setHeaderTitle("空气系统设置");
```

```
        menu1.add(0, SUB_MENU_UPLOAD_A, 0, "设置参数 A");
        menu1.add(0, SUB_MENU_UPLOAD_B, 0, "设置参数 B");
        SubMenu menu2 = (SubMenu) menu.addSubMenu(0, MENU_02, 1,"城市温度");
        menu2.setHeaderTitle("温度系统设置");
        menu2.add(0, SUB_MENU_UPLOAD_C, 0, "设置参数 C");
        menu2.add(0, SUB_MENU_UPLOAD_D, 0, "设置参数 D");
        return true;
    }
    //当被菜单选项被选中时调用 onOptionsItemSelected()方法
    @Override
    public boolean onOptionsItemSelected(MenuItem item) {
        switch (item.getItemId()) {//通过该方法获取菜单选项 ID
            case SUB_MENU_UPLOAD_A:
                Menu1Counter++;
                return true;
            case SUB_MENU_UPLOAD_B:
                Menu2Counter++;
                return true;
        }
        return false;
    }
}
```

9.5　任务验证

在 Android Studio 开发环境中打开本项目的例程，编译通过后运行程序，如图 9.7 所示。Android 新的模拟器不支持在界面上调出菜单，可以通过快捷键 Ctrl+M 调出菜单，如图 9.8 所示。

图 9.7　程序运行效果

图9.8　在界面中调出菜单

9.6　开发小结

本任务主要介绍Android的菜单。通过本任务的学习，读者可以掌握Android菜单的使用方法，并在城市环境系统中添加功能菜单。

9.7　思考与拓展

（1）Android的选项菜单和上下文菜单有什么区别?

（2）请尝试在本任务的主菜单中添加一个“报警系统”子菜单。

任务 10

城市灯光控制系统界面事件的处理

本任务介绍 Android 系统界面事件，主要内容包括 Android 界面事件和监听器、Android 按键事件的处理，以及 Android 屏幕触摸事件的处理等内容。通过本任务的学习，读者可以掌握 Android 系统界面事件的处理方法，实现城市灯光控制系统界面事件的处理。

10.1 开发场景：如何用 Android 开发一个城市灯光控制系统界面

随着科技的不断进步，用手机控制家用电器已经不再是科幻电影中的情节了。能不能开发一个简单的 Android 应用来控制灯光呢？本任务通过 Android 的按键事件和触摸事件来实现灯光的控制。

10.2 开发目标

（1）知识目标：熟悉 Android 按键事件和屏幕触摸事件的处理机制。

（2）技能目标：掌握 Android 系统界面事件的处理方法。

（3）任务目标：掌握 Android 按键事件和屏幕触摸事件的处理方法，实现城市灯光控制系统界面事件的处理。

10.3 原理学习：Android 系统界面事件实现

10.3.1 监听器

单击事件是 Android 控件开发中最常见的事件之一，可以通过添加监听器来监听该事件。添加监听器的方式有以下两种：

（1）直接在 onCreate( )方法中添加监听器，使用匿名内部类作为监听器类。

```
public class MainActivity extends AppCompatActivity {
    private Button btnOk;
```

```
    @Override
    protected void onCreate(Bundle savedInstanceState) {
        super.onCreate(savedInstanceState);
        setContentView(R.layout.activity_main);
        Button btnOk = findViewById(R.id.btn_ok);
        btnOk.setOnClickListener(new View.OnClickListener() {
            @Override
            public void onClick(View v) {
                //此处添加监听器
                Toast.makeText(MainActivity.this, "您单击了按钮", Toast.LENGTH_SHORT).show();
            }
        });
    }
}
```

在这里为按钮单击事件添加监听器，在单击按钮时会弹出提示框。

（2）使用接口方式注册监听器。

```
public class MainActivity extends AppCompatActivity implements View.OnClickListener {
    private Button btnOk;
    @Override
    protected void onCreate(Bundle savedInstanceState) {
        super.onCreate(savedInstanceState);
        setContentView(R.layout.activity_main);
        Button btnOk = findViewById(R.id.btn_ok);
        btnOk.setOnClickListener(this);
    }
    @Override
    public void onClick(View v) {
        switch (v.getId()) {
            case R.id.btn_ok:
                Toast.makeText(MainActivity.this, "您单击了按钮", Toast.LENGTH_SHORT).show();
                break;
            default:
                break;
        }
    }
}
```

10.3.2 Android 的界面事件和监听器

Android 中的界面事件按类型可以分为按键事件和屏幕触摸事件，控制器根据界面事件（UI Event）类型的不同，将界面事件传递给界面控件中不同的函数进行处理。

（1）按键事件（KeyEvent）：将事件传递给 onKey()函数进行处理。

（2）触摸事件（TouchEvent）：将事件传递给 onTouch()函数进行处理。

Android 界面事件的传递与处理应遵循如下规则。如果界面控件设置了事件监听器，则会优先将界面事件传递给事件监听器；如果界面控件没有设置事件监听器，界面事件则会直接传递给界面控件的其他事件处理函数。即使界面控件设置了事件监听器，界面事件也可以再次传递给其他事件处理函数，是否继续将界面事件传递给其他事件处理函数由事件监听器处理函数的返回值决定。如果事件监听器处理函数的返回值为“真”，表示该事件已经完成处理过程，则不需要其他事件处理函数参与处理过程，这样就不会再继续传递界面事件；如果事件监听器处理函数的返回值为“假”，表示该界面事件没有完成处理过程，或需要其他事件处理函数参与处理过程，这时就会将界面事件传递给其他事件处理函数。

10.3.3　Android 按键事件的处理

Android 按键事件处理主要是针对 View 和 Activity 进行的。按键事件的处理如下：

（1）如果 View 没有获得焦点，则将按键事件传递给 Activity 处理。

（2）如果 View 获得焦点，则按键事件会优先传递到 View 的回调方法。如果 View 的回调方法返回“假”，则按键事件继续传递到 Activity 处理；反之，按键事件不会继续传递。

View.setFocusable(true)方法用来获得焦点；public boolean onKeyDown(int keyCode, KeyEvent msg)方法用来处理按键按下事件；public boolean onKeyUp(int keyCode, KeyEvent msg)方法用来处理按键抬起事件。

注意：

（1）要响应按键事件，就需要在构造函数中调用 this.setFocusable(true)。

（2）按键事件的 onKeyDown 和 onKeyUp 是相互独立的，不会相互影响。

（3）无论在 View 还是在 Activity 中，建议在覆写事件回调方法时，只对处理过的按键事件返回 true，没有处理过的按键事件应调用其父类方法，否则，其他未处理的事件不会被传递到合适的目标组件中，如按键失效问题。

为了处理 Android 的按键事件，先需要设置按键事件监听器，并重载 onKey()函数。Android 按键事件的监听及事件传递给处理函数的示例如下：

```
entryText.setOnKeyListener(new OnKeyListener(){
    @Override
    public boolean onKey(View view, int keyCode, KeyEvent keyEvent){
        //处理过程代码
        return true/false;
    }
}
```

第 1 行代码用于设置控件的按键事件监听器。

第 3 行代码的 onKey()函数的参数如下：view 表示产生按键事件的界面控件；keyCode 表示按键编码；keyEvent 包含事件的详细信息，如按键的重复次数、编码和标志等。

第 5 行代码是 onKey()函数的返回值。当返回 true 时，阻止按键事件的传递；当返回 false 时，允许继续传递按键事件。

10.3.4 Android 屏幕触摸事件的处理

在 Android 系统中，TouchEvent 是基础事件之一。例如，在多用户界面嵌套的应用中：如果触摸的用户界面部分没有重叠，则该屏幕触摸事件只属于某个单独的用户界面，那么只有这个单独的用户界面能够捕获到触摸事件；如果触摸的是用户界面重叠的部分，那么首先捕获到屏幕触摸事件的是父类 View，父类 View 再根据特定方法的返回值来决定屏幕触摸事件的处理者。

在 Android 中，每个 View 的子类都有 3 个和屏幕触摸事件处理密切相关的方法，分别如下：

```
public boolean dispatchTouchEvent(MotionEvent ev);            //分发屏幕触摸事件
public boolean onInterceptTouchEvent(MotionEvent ev);         //拦截屏幕触摸事件
public boolean onTouchEvent(MotionEvent ev);                  //处理屏幕触摸事件
```

其中，onTouchEvent()方法定义在 View 类中，当发生屏幕触摸事件时，首先将屏幕触摸事件传递到该 View 类，由该 View 类进行处理时，onTouchEvent()方法将会被执行。屏幕触摸事件的处理流程如图 10.1 所示。

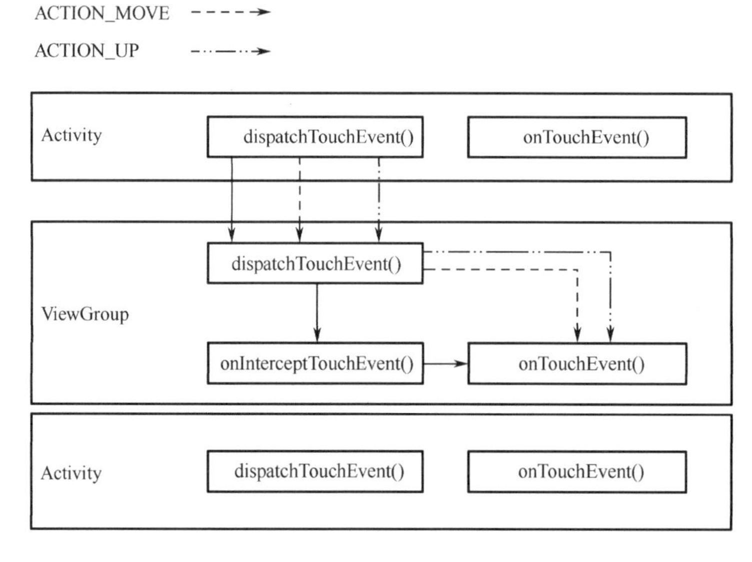

图 10.1 屏幕触摸事件的处理流程

dispatchTouchEvent()和 onInterceptTouchEvent()这两个方法是在 ViewGroup 中定义的，因为只有 ViewGroup 才包含子类 View 和子类 ViewGroup，所以在多用户界面嵌套时需要通过上述的两个方法来决定是否监听处理连续的屏幕触摸动作和屏幕触摸动作由谁去截获处理。

（1）dispatchTouchEvent()方法：如果返回值为 false，表示捕获到一个屏幕触摸事件，则 View 类便会调用 onInterceptTouchEvent()方法进行处理；如果返回值为 true，则 View 类将监听处理一连串的屏幕触摸事件。

（2）onInterceptTouchEvent()方法：如果返回值为 false，View 类将不处理传递过来的屏幕触摸事件，而把屏幕触摸事件传递给子类 View；如果返回值为 true，该 View 类将把屏幕触摸事件截获并进行处理，不会把屏幕触摸事件传递给子类 View。

（3）onTouchEvent()方法：如果返回值为 true，表示屏幕触摸事件处理完毕，将等待下一次的屏幕触摸事件；如果返回值为 false，则会返回调用用户界面重叠部分的处于上层的相邻 View 类的 onTouchEvent()方法。如果没有用户界面重叠的相邻 View 类，将返回调用父类 View 的 onTouchEvent()方法。如果到了最外层的父类 View 的 onTouchEvent()方法还是返回 false，则 TouchEvent 消失。

当发生屏幕触摸事件时，首先 Activity 将屏幕触摸事件传递给最顶层的 View 类，屏幕触摸事件最先到达最顶层 View 类的 dispatchTouchEvent()方法，然后由 dispatchTouchEvent()方法进行分发。如果 dispatchTouchEvent()方法返回 true，则交给 View 类的 onTouchEvent()方法处理；如果 dispatchTouchEvent()方法返回 false，则交给 View 的 interceptTouchEvent()方法来决定是否要拦截这个事件。

覆写 Activity 的 onTouchEvent()方法后，当发生屏幕触摸事件时，此方法就会被调用。覆写 onTouchEvent()方法的代码如下：

```
@Override
    public boolean onTouchEvent(MotionEvent event) {
        switch (event.getAction()) {
            case MotionEvent.ACTION_DOWN://0
                Log.e("TAG", "MainActivity onTouchEvent 按下");
                break;
            case MotionEvent.ACTION_UP://1
                Log.e("TAG", "MainActivity onTouchEvent 抬起");
                break;
            case MotionEvent.ACTION_MOVE://2
                Log.e("TAG", "MainActivity onTouchEvent 移动");
                break;
        }
        return super.onTouchEvent(event);
    }
```

10.4　开发实践：城市灯光控制系统界面事件的处理

10.4.1　开发设计

根据城市灯光控制系统的需求，使用 Switch 控件作为系统的控制按钮，使用 SeekBar 控件控制灯光的亮度。城市灯光控制系统界面如图 10.2 所示，界面布局结构如图 10.3 所示。

本项目基于 Android 的 Empty Activity 模板创建，主要编写 MainActivity.java 文件与 activity_main.xml 文件，项目目录结构如图 10.4 所示。

图 10.2　城市灯光控制系统界面

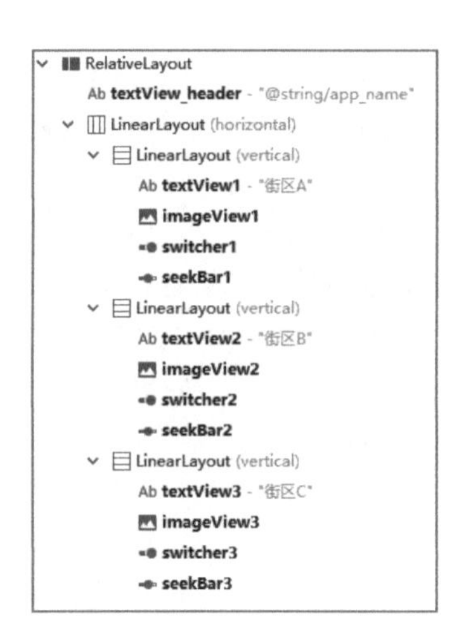

图 10.3　界面布局结构

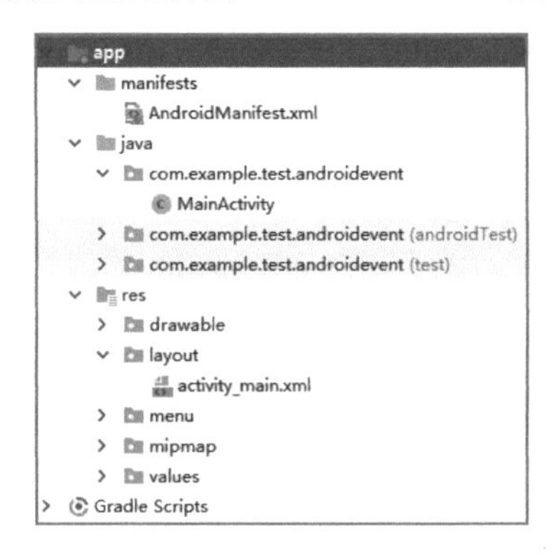

图 10.4　项目目录结构

项目开发步骤如下：

（1）编写 activity_main.xml 布局文件，设置界面上 TextView、ImageView、Switcher、SeekBar 等控件的布局位置。

（2）在 MainActivity.java 文件中实现类的 init(final ImageView imageView, Switcher switcher, SeekBar seekBar)方法。

（3）在 init()方法中设置 ImageView、Switcher、SeekBar 等控件的状态。

（4）在 init()方法中设置 switcher 对象的 setOnCheckedChangeListener 监听器，当用户单击按钮时通过 Toast 来提示开关信息。

（5）在 onCreate()方法中实例化 TextView、ImageView、Switcher、SeekBar 控件的对象，并通过调用 init(imageView, switch, seekBar)方法进行初始化。

10.4.2　功能实现

1. 界面设计

```
<RelativeLayout xmlns:android="http://schemas.android.com/apk/res/android"
    xmlns:app="http://schemas.android.com/apk/res-auto"
```

```
xmlns:tools="http://schemas.android.com/tools"
android:layout_width="match_parent"
android:layout_height="match_parent">
<TextView
    android:id="@+id/textView_header"
    android:layout_width="match_parent"
    android:layout_height="wrap_content"
    android:layout_alignParentStart="true"
    android:layout_alignParentTop="true"
    android:layout_marginStart="0dp"
    android:layout_marginTop="10dp"
    android:layout_marginBottom="10dp"
    android:gravity="center"
    android:textSize="24sp"
    android:text="@string/app_name" />
<LinearLayout
    android:layout_width="match_parent"
    android:layout_height="match_parent"
    android:layout_alignParentStart="true"
    android:layout_centerVertical="true"
    android:layout_below="@+id/textView_header"
    android:orientation="horizontal"
    android:layout_marginRight="80dp">
    <LinearLayout
        android:layout_width="match_parent"
        android:layout_height="match_parent"
        android:layout_weight="1"
        android:orientation="vertical">
        <TextView
            android:id="@+id/textView1"
            android:layout_width="wrap_content"
            android:layout_height="wrap_content"
            android:text="街区 A" />
        <ImageView
            android:id="@+id/imageView1"
            android:layout_width="180dp"
            android:layout_height="160dp"
            android:scaleType="fitXY"
            android:src="@drawable/jie1" />
        <Switch
            android:id="@+id/switcher1"
            android:layout_width="wrap_content"
            android:layout_height="wrap_content"
            android:text="" />
        <SeekBar
            android:id="@+id/seekBar1"
            android:layout_width="180dp"
```

```
            android:layout_height="wrap_content" />
    </LinearLayout>
    <LinearLayout
        android:layout_width="match_parent"
        android:layout_height="match_parent"
        android:layout_weight="1"
        android:orientation="vertical">
        <TextView
            android:id="@+id/textView2"
            android:layout_width="wrap_content"
            android:layout_height="wrap_content"
            android:text="街区 B" />
        <ImageView
            android:id="@+id/imageView2"
            android:layout_width="180dp"
            android:layout_height="160dp"
            android:scaleType="fitXY"
            android:src="@drawable/jie2" />
        <Switch
            android:id="@+id/switcher2"
            android:layout_width="wrap_content"
            android:layout_height="wrap_content"
            android:text="" />
        <SeekBar
            android:id="@+id/seekBar2"
            android:layout_width="180dp"
            android:layout_height="wrap_content" />
    </LinearLayout>
    <LinearLayout
        android:layout_width="match_parent"
        android:layout_height="match_parent"
        android:layout_weight="1"
        android:orientation="vertical">
        <TextView
            android:id="@+id/textView3"
            android:layout_width="wrap_content"
            android:layout_height="wrap_content"
            android:text="街区 C" />
        <ImageView
            android:id="@+id/imageView3"
            android:layout_width="180dp"
            android:layout_height="160dp"
            android:scaleType="fitXY"
            android:src="@drawable/jie3" />
        <Switch
            android:id="@+id/switcher3"
            android:layout_width="wrap_content"
```

```
                android:layout_height="wrap_content"
                android:text="" />
            <SeekBar
                android:id="@+id/seekBar3"
                android:layout_width="180dp"
                android:layout_height="wrap_content" />
        </LinearLayout>
    </LinearLayout>
</RelativeLayout>
```

2. 主界面设计

主界面设计主要是修改 MainActivity.java 文件，代码如下：

```
public class MainActivity extends Activity {
    @Override
    protected void onCreate(Bundle savedInstanceState) {
        super.onCreate(savedInstanceState);
        //去掉 TitleBar
        requestWindowFeature(Window.FEATURE_NO_TITLE);
        setContentView(R.layout.activity_light);
        ImageView imageView1 = findViewById(R.id.imageView1);
        Switch switch1 = findViewById(R.id.switcher1);
        SeekBar seekBar1 = findViewById(R.id.seekBar1);
        ImageView imageView2 = findViewById(R.id.imageView2);
        Switch switch2 = findViewById(R.id.switcher2);
        SeekBar seekBar2 = findViewById(R.id.seekBar2);
        ImageView imageView3 = findViewById(R.id.imageView3);
        Switch switch3 = findViewById(R.id.switcher3);
        SeekBar seekBar3 = findViewById(R.id.seekBar3);
        init(imageView1, switch1, seekBar1);
        init(imageView2, switch2, seekBar2);
        init(imageView3, switch3, seekBar3);
    }
    private void init(final ImageView imageView, Switch switcher, SeekBar seekBar) {
       imageView.setAlpha(1.0f);
        switcher.setChecked(true);
        seekBar.setProgress(50);
        switcher.setOnCheckedChangeListener(new CompoundButton.OnCheckedChangeListener() {
            @Override
            public void onCheckedChanged(CompoundButton compoundButton, boolean b) {
                if(b) {
                    Toast.makeText(MainActivity.this, "打开街区灯光。", Toast.LENGTH_SHORT).show();
                    imageView.setAlpha(1.0f);
                } else {
                    Toast.makeText(MainActivity.this, "关闭街区灯光。", Toast.LENGTH_SHORT).show();
                    imageView.setAlpha(0.3f);
                }
```

```
                }
            });
            switcher.setOnTouchListener(new View.OnTouchListener() {
                @Override
                public boolean onTouch(View v, MotionEvent event) {
                    if (event.getAction()==MotionEvent.ACTION_DOWN){
                        Log.i("onTouch","按下");
                    }else if (event.getAction()==MotionEvent.ACTION_UP){
                        Log.i("onTouch","抬起");
                    }
                    return false;
                }
            });
        }
        @Override
        public boolean onCreateOptionsMenu(Menu menu) {
            //Inflate the menu; this adds items to the action bar if it is present.
            getMenuInflater().inflate(R.menu.main, menu);
            return true;
        }
    }
```

10.5 任务验证

在 Android Studio 开发环境中打开本项目的例程，编译通过后运行程序，如图 10.5 所示。

图 10.5 程序运行效果

单击“打开”或“关闭”按钮时，系统会弹出消息提示，单击“关闭”按钮时图片会变灰，单击“打开”按钮时图片会变亮，如图 10.6 所示。

图 10.6　单击“打开”或“关闭”按钮后的效果

10.6　开发小结

本任务主要介绍 Android 的按键事件（KeyEvent）及屏幕触摸事件（TouchEvent）。通过本任务的学习，读者可以掌握 Android 界面事件的处理方法，实现城市灯光控制系统界面事件的处理。

10.7　思考与拓展

（1）在 Android 的屏幕触摸事件中，ACTION_DOWN 和 ACTION_UP 分别代表什么含义？

（2）请尝试增加灯光亮度控制功能，通过拖动 SeekBar 控件来调节灯光的亮度。

第 3 篇

Android 开发进阶

本篇主要介绍 Android 的高级知识和语法，通过开发实践帮助读者熟悉 Android 的开发进阶。本篇共有 7 个任务：

任务 11 为工厂通风系统界面的切换。

任务 12 为工厂火警监测系统界面的设计。

任务 13 为设备列表管理界面的设计。

任务 14 为智能电表日志的记录。

任务 15 为光照度记录的查询。

任务 16 为智能医疗仪表图形的动态显示。

任务 17 为远程控制服务端的通信。

任务 11

工厂通风系统界面的切换

本任务介绍 Android 中的 Activity 和 Service，主要内容包括 Activity 的生命周期与创建、Activity 间的数据传递与交互、Service 与 Thread 线程，以及 IntentService 的使用。通过本任务的学习，读者可以掌握 Android 系统的 ActivityAction 应用，实现工厂通风系统界面的切换。

11.1 开发场景：如何实现工厂通风系统界面的切换

假设要实现一个工厂通风系统，要求设计一个主界面和一个子界面，在主界面中显示工厂通风系统的各项参数，在子界面中由人工采集工厂通风系统的数据，数据采集完成把数据传递到主界面。本任务通过 Activity 之间数据传递与交互等功能来实现工厂通风系统界面的切换。

11.2 开发目标

（1）知识目标：熟悉 Activity 的生命周期与创建、Activity 间的数据传递与交互、Service 与 Thread 线程、IntentService 的使用。

（2）技能目标：掌握 Android 系统中 ActivityAction 的应用方法。

（3）任务目标：学习 Activity 的生命周期与创建、Activity 间的数据传递与交互、Service 与 Thread 线程、IntentService 的使用，掌握 Android 系统中 ActivityAction 的应用方法，实现工厂通风系统界面的切换。

11.3 原理学习：Android 中的 Activity 和 Service

11.3.1 Android 中的 Activity

1．Activity 的生命周期与创建

Activity 是应用与用户交互的接口，它为用户完成相关操作提供了一个界面。当创建

Activity 后，可以通过调用 setContentView(View)方法来给创建的 Activity 指定一个布局界面，这个界面就是与用户交互的接口。

Android 是通过 Activity 栈的方式来管理 Activity 的，而 Activity 本身则是通过生命周期的方法来管理的自己的创建与销毁的。Activity 生命周期流程如图 11.1 所示。

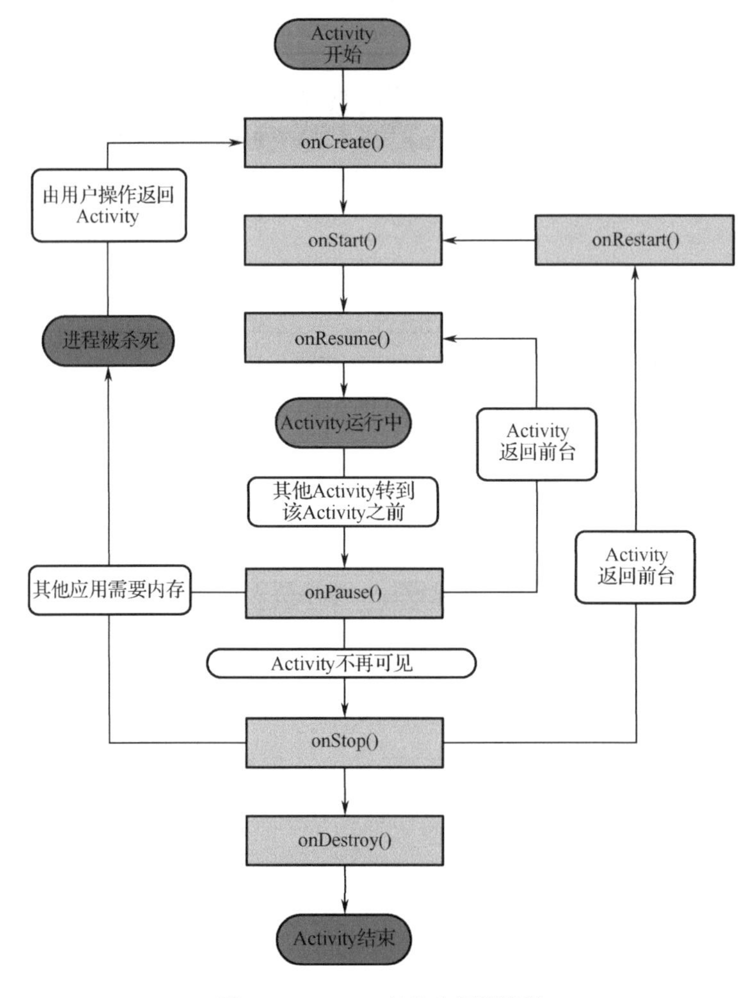

图 11.1　Activity 的生命周期流程

Activity 的状态有以下几种：

（1）Active/Running：当 Activity 处于 Active/Running 状态时，此时 Activity 处于栈顶，是可见状态，可与用户进行交互。

（2）Paused：当 Activity 失去焦点，被一个新的非全屏的 Activity 或一个透明的 Activity 放置在栈顶时，该 Activity 就转化为 Paused 状态。此时该 Activity 失去了与用户交互的能力，但其所有的状态信息及其成员变量都还存在，在系统内存不足的情况下，才有可能被系统回收掉。

（3）Stopped：当一个 Activity 被另一个 Activity 完全覆写时，被覆写的 Activity 就会进入 Stopped 状态，此时它不再可见，但同样保存着其所有状态信息及其成员变量。

（4）Killed：当 Activity 被系统回收掉时，Activity 就处于 Killed 状态。

生命周期指在有用户参与的情况下，Activity 经历从创建、运行、停止到销毁等正常的生命周期过程。在生命周期中经常调用的方法如下：

（1）onCreate()：在创建 Activity 时调用该方法，它是 Activity 生命周期内第一个被调用的方法。在创建 Activity 时需要覆写 onCreate()方法，然后在该方法中进行一些初始化的操作，如通过 setContentView()方法设置界面布局的资源、初始化所需要的组件信息等。

（2）onStart()：此方法被调用时表示 Activity 正在启动，此时 Activity 处于可见状态，只是还没有在前台显示，因此无法与用户进行交互。

（3）onResume()：当此方法调用时表示 Activity 已在前台可见，可与用户进行交互（处于 Active/Running 状态）。onResume()方法与 onStart()方法的相同点是两者都表示 Activity 可见，只不过在调用 onStart()方法时 Activity 处于后台，无法与用户交互，而在调用 onResume()方法时 Activity 在前台，可与用户交互。从图 11.1 也可以看出，当 Activity 停止后（onPause()方法和 onStop()方法被调用），在重新回到前台时也会调用 onResume()方法，因此也可以在 onResume()方法中初始化一些资源，例如重新初始化在 onPause()方法或者 onStop()方法中释放的资源。

（4）onPause()：此方法被调用时表示 Activity 正在停止（Paused）状态，一般情况下会接着调用 onStop()方法。但通过流程图还可以看到一种情况，即 onPause()方法执行后直接调用了 onResume()方法，这是因为用户操作使当前 Activity 退居后台后又迅速地再回到前台，此时就会调用 onResume()方法。在 onPause()方法中可以进行一些数据存储、动画停止或资源回收等操作，但不能太耗时，因为这可能会影响到新 Activity 的显示。onPause()方法执行完成后，新 Activity 的 onResume()方法才会被执行。

（5）onStop()：一般在 onPause()方法执行完成直接调用该方法，表示 Activity 即将停止或者完全被覆写（处于 Stopped 状态），此时 Activity 不可见，仅在后台运行。在 onStop()方法中可以释放资源。

（6）onRestart()：表示 Activity 正在重新启动，当 Activity 由不可见变为可见时，将调用该方法。例如，当用户打开一个新 Activity 时，当前的 Activity 就会被暂停（调用 onPause()方法和 onStop()方法）；当返回当前 Activity 时，就会调用 onRestart()方法。

（7）onDestroy()：此时 Activity 正在被销毁，也是在生命周期内最后调用的方法，可以在该方法中进行一些回收工作，如最终释放资源。

下面验证上述流程的几种方法，代码如下：

```
public class MainActivity extends AppCompatActivity {
    private Button btn;
    //在 Activity 创建时调用 onCreate()方法
    @Override
    protected void onCreate(Bundle savedInstanceState) {
        super.onCreate(savedInstanceState);
        setContentView(R.layout.activity_main);
        Log.e("MainActivity", "onCreate is invoke!!!");
        btn = findViewById(R.id.btn);
        btn.setOnClickListener(new View.OnClickListener() {
            @Override
            public void onClick(View v) {
```

```
                Intent i = new Intent(MainActivity.this, SecondActivity.class);
                startActivity(i);
            }
        });
    }
    //当 Activity 从后台重新回到前台时调用 onRestart()方法
    @Override
    protected void onRestart() {
        super.onRestart();
        Log.e("MainActivity", "onRestart is invoke!!!");
    }
    //当创建 Activity 或者 Activity 从后台重新回到前台时调用 onStart()方法
    @Override
    protected void onStart() {
        super.onStart();
        Log.e("MainActivity", "onStart is invoke!!!");
    }
    //当 Activity 被创建或者从被覆写、后台重新回到前台时调用 onResume()方法
    @Override
    protected void onResume() {
        super.onResume();
        Log.e("MainActivity", "onResume is invoke!!!");
    }
    //当 Activity 被覆写到后台或者锁屏时调用 onPause()方法
    @Override
    protected void onPause() {
        super.onPause();
        Log.e("MainActivity", "onPause is invoke!!!");
    }
    //当退出当前 Activity 或者跳转到新 Activity 时调用 onStop()方法
    @Override
    protected void onStop() {
        super.onStop();
        Log.e("MainActivity", "onStop is invoke!!!");
    }
    //当退出当前 Activity 时调用 onDestroy()方法，调用该方法之后 Activity 就结束了
    @Override
    protected void onDestroy() {
        super.onDestroy();
        Log.e("MainActivity", "onDestroy is invoke!!!");
    }
}
```

Activity 的生命周期日志如图 11.2 所示。

运行 Activity 后，Activity 的生命周期日志首先显示关于 onCreate()、onStart()和 onResume()等方法的信息，此时 MainActivity 处于正常运行状态。当单击按钮打开 SecondActivity 时，

MainActivity 先后调用 onPause()方法和 onStop()方法。当从 SecondActivity 返回时，MainActivity 先后调用 onRestart()方法、onStart()方法和 onResume()方法，此时 MainActivity 恢复到正常运行状态。当单击返回按钮时，MainActivity 调用 onPause()方法、onStop()方法后调用 onDestroy()方法完成 Activity 的销毁。

```
09:26:56.796 9995-9995/com.example.myapplication E/MainActivity: onCreate is invoke!!!
09:26:56.820 9995-9995/com.example.myapplication E/MainActivity: onStart is invoke!!!
09:26:56.836 9995-9995/com.example.myapplication E/MainActivity: onResume is invoke!!!
09:27:09.649 9995-9995/com.example.myapplication E/MainActivity: onPause is invoke!!!
09:27:10.218 9995-9995/com.example.myapplication E/MainActivity: onStop is invoke!!!
09:27:15.580 9995-9995/com.example.myapplication E/MainActivity: onRestart is invoke!!!
09:27:15.581 9995-9995/com.example.myapplication E/MainActivity: onStart is invoke!!!
09:27:15.589 9995-9995/com.example.myapplication E/MainActivity: onResume is invoke!!!
09:27:19.555 9995-9995/com.example.myapplication E/MainActivity: onPause is invoke!!!
09:27:20.048 9995-9995/com.example.myapplication E/MainActivity: onStop is invoke!!!
09:27:20.054 9995-9995/com.example.myapplication E/MainActivity: onDestroy is invoke!!!
```

图 11.2　Activity 的生命周期日志

2．Activity 间的数据传递与交互

（1）数据传递。假设有两个 Activity，即 MainActivity 和 SecondActivity，MainActivity 中有一个字符串，现在想把这个字符串传递到 SecondActivity，可以使用 Intent 的 putExtra()方法和 getStringExtra()方法在 Activity 之间传递数据。

MainActivity 需要将数据传递给 SecondActivity，代码如下：

```
//MainActivity（发送者）
    button.setOnClickListener(new View.OnClickListener() {
        @Override
        public void onClick(View v) {
            String dataStr = "数据传递: Hello";
            Intent intent = new Intent(MainActivity.this, SecondActivity.class);
            intent .putExtra("ExtData",dataStr );
            startActivity(intent );
        }
});
//SecondActivity（接收者）
public class SecondActivity extends Activity {
    @Override
    protected void onCreate(Bundle savedInstanceState){
        super.onCreate(savedInstanceState);
        setContentView(R.layout.secondLayout);
        //通过 Intent 获取数据
        Intent intent = getIntent();
        String dataStr = intent .getStringExtra("ExtData");
        Toast.makeText(SecondActivity.this,dataStr ,Toast.LENGTH_SHORT).show();
    }
}
```

（2）数据返回。数据一般是在 Activity 的销毁时返回的。假设现在需要从 SecondActivity 返回数据给 MainActivity，代码如下：

```
//SecondActivity
public class SecondActivity extends Activity {
    @Override
    protected void onCreate(Bundle savedInstanceState){
        super.onCreate(savedInstanceState);
        setContentView(R.layout.secondLayout);
        Button butRet= findViewById(R.id.buttonRet);
        butRet.setOnClickListener(new View.OnClickListener() {
            @Override
            public void onClick(View v) {
                Intent intent = new Intent();
                intent.putExtra("dataRet", "数据返回:123…");
                //绑定 result_code 和 mIntent 内容
                setResult(200, intent);
                finish();
            }
        });
    }
}
```

MainActivity 接收数据时需要三个参数，分别是请求码 requestCode、结果码 resultCode 和 data。

```
public class MainActivity extends AppCompatActivity {
    @Override
    protected void onCreate(Bundle savedInstanceState) {
        super.onCreate(savedInstanceState);
        setContentView(R.layout.activity_main);
        Button butReq = findViewById(R.id.buttonReq);
        butReq .setOnClickListener(new View.OnClickListener() {
            @Override
            public void onClick(View v) {
                Intent intent = new Intent(MainActivity.this, SecondActivity.class);
                startActivityForResult(intent, 100);
            }
        });
    }
    @Override
    protected void onActivityResult(int requestCode, int resultCode, Intent data){
        if (requestCode == 100 && requestCode==200 ){
            String retData = data.getStringExtra("dataReturn");
            Toast.makeText(MainActivity .this, retData ,Toast.LENGTH_SHORT).show();
        }
    }
}
```

（3）在 Bundle 中传递和保存数据。如果当前的 Activity 在销毁时需要保存一些临时数据，则可以通过 onSaveInstanceState()函数来保存这些临时数据。

① 保存数据的代码如下：

```
@Override
    protected void onSaveInstanceState(Bundle outState){
        super.onSaveInstanceState(outState);
        String testData = "Test Data…";
        outState.putString("dataKey",testData);
    }
```

② 取出数据的代码如下：

```
public class ActivityLifeCycleTest extends AppCompatActivity {
    public static final String TAG = "MainActivity";
    @Override
    protected void onCreate(Bundle savedInstanceState) {
        super.onCreate(savedInstanceState);
        setContentView(R.layout.activityTest);
        if (savedInstanceState != null){
            String dataStr = savedInstanceState.getString("dataKey");
            Log.d(TAG, "dataKey = " + dataStr);
        }
        ……
    }
}
```

11.3.2　Android 中的 Service

1．Service 与 Thread

（1）Service 与 Thread 的定义。

① Service（服务）：Service 是 Android 四大组件之一，适合执行不需要与用户交互且要求长期执行的任务。需要注意的是，Service 依赖于创建 Service 时所在的应用程序进程。

② Thread（线程）：Thread 是程序执行的最小单元，可以用线程来执行一些异步操作。

（2）Service 与 Thread 的使用。

① 线程的使用。通常会使用线程来进行耗时操作。例如，有时需要根据任务的执行结果来更新显示相应的 UI 控件，Android 的 UI 控件与其他许多 GUI 库一样，线程是不安全的，必须在主线程中操作，因此需要采用异步消息机制，即使用基于异步消息机制的 AsyncTask 抽象类在 doInBackground()方法（运行在子线程中）执行耗时操作，在 onProgressUpdate()方法（运行在主线程中）中进行 UI 操作，在 onPostExecute()方法（运行在主线程中）中执行任务的收尾工作。

```
private class MyTask extends AsyncTask<Params, Progress, Result> {
    //执行线程任务前的操作
    @Override
    protected void onPreExecute() {
    }
```

```
        //接收输入参数、执行任务中的耗时操作、返回线程任务执行的结果
        @Override
        protected String doInBackground(String params) {
            //自定义的线程任务
        }
        //在主线程中显示任务执行的进度
        @Override
        protected void onProgressUpdate(Integer progresses) {
        }
        //接收任务执行结果，将执行结果显示到 UI 控件
        @Override
        protected void onPostExecute(String result) {
            //UI 操作
        }
    }

    //创建 AsyncTask 抽象类的实例对象
    MyTask mTask = new MyTask();
    mTask.execute();
```

② Service 的使用。主线程一直处于运行状态，如果直接在 Service 中进行耗时操作，则可能会阻塞主线程，因此需要在 Service 中手动创建子线程，在子线程中进行耗时操作。一个典型的 Service 如下：

```
public class MyService extends Service {
    ……
    @Override
    public int onStartCommand(Intent intent,int flags,int startId)
    {
        new Thread（new Runnable()
        {
            public void run()
            {
                //处理具体逻辑
            }
        }).start();
        return super.onStartCommand（intent，flags，startId）;
    }
}
```

Service 一旦启动，必须调用 stopSelf()方法或 stopService()方法才能停止。为了创建一个异步自动停止的 Service，Android 专门提供了一个 IntentService 类，只需要新建一个 MyIntentService 类（该类继承自 IntentService 类），在 onHandlerIntent()方法中执行耗时操作即可，该方法在子线程中运行，且在运行结束后会自动停止。

2．IntentService 的使用

Android 中的 IntentService 类继承自 Service 类，Service 类的回调方法，如 onCreate()、 onStartCommand()、onBind()、onDestroy()等，都运行在主线程中。当通过 startService()方法启动 Service 之后，需要在 Service 的 onStartCommand()方法中完成工作，但是 onStartCommand()方法运行在主线程中，如果需要在此处完成一些网络请求或 I/O 等耗时操作，就可能会阻塞主线程。为了解决这个问题，可以在 onStartCommand()方法中创建一个新线程，把耗时的操作放到这个新线程中进行，将在 onStartCommand()方法中开启的新线程作为工作线程来进行耗时操作。

为了简化开发带有工作线程的 Service，Android 开发了一个类——IntentService，IntentService 的特点如下：

（1）IntentService 自带一个工作线程，当需要在 Service 进行可能阻塞主线程的操作时，可以使用 IntentService。

（2）将需要实现的主要功能放入 IntentService 中的 onHandleIntent()方法中，当通过 startService(intent)方法启动 IntentService 后，Android Framework 最终会调用 onHandleIntent()方法，并将 intent 传入该方法，onHandleIntent()方法运行在 IntentService 的工作线程中。

（3）当通过 startService()方法多次启动 IntentService 时，就会产生多个任务。由于 IntentService 只有一个工作线程，所以 onHandleIntent()方法每次只能处理一个任务。面对多个任务，IntentService 该如何处理呢？IntentService 是按照先后顺序处理任务的，先将 intent1 传入 onHandleIntent()方法，让其完成任务 1，再将 intent2 传入 onHandleIntent()方法，让其完成任务 2……，直至所有任务完成。IntentService 不能并行地处理多个任务，当所有的任务都完成时，调用 onDestroy()方法销毁 IntentService。

IntentService 继承自 Service 类，IntentService 不仅覆写了 onCreate()、onStartCommand()、onStart()、onDestroy()等方法，还添加了 onHandleIntent()方法，这几个方法分析如下。

（1）onCreate()：在 onCreate()方法中，利用 mName 作为线程名称，创建 HandlerThread，HandlerThread 是 IntentService 的工作线程。HandlerThread 在执行了 onStart()方法之后就关联了消息队列和 Looper，并且消息队列开始循环。

（2）onStartCommand()：IntentService 覆写了 onStartCommand()方法，在该方法内部调用 onStart()方法。

（3）onStart()：在 onStart()方法中创建 Message 对象，并将 Intent 对象作为 Message 的 obj 参数，这样 Message 与 Intent 就关联起来了，然后通过 HandlerThread 的 sendMessage()方法将关联了 Intent 的 Message 发送给 HandlerThread。

（4）onHandleIntent()：onStart()方法通过 sendMessage()方法将 Message 放入 HandlerThread 所关联的消息队列中后，HandlerThread 所关联的 Looper 对象会从消息队列中取出一个 Message，然后将其传入 HandlerThread 的 handleMessage()方法中。在 handleMessage()方法中首先通过 Message 的 obj 获取原始的 Intent 对象，然后将其作为参数传给 onHandleIntent()方法让其执行。handleMessage()方法运行在 HandlerThread 中，所以 onHandleIntent()方法也运行在 HandlerThread 中。在执行完了 onHandleIntent()方法之后，需要调用 stopSelf(startId)方法来声明某个任务完成了。当完成所有的任务时，就调用 onDestroy()方法来销毁 IntentService。

（5）onDestroy()：在完成所有的任务时，Service 会被销毁并调用 onDestroy()方法，以及 HandlerThread 的 quit()方法（该方法终止消息循环）。

总结：IntentService 可以在线程中完成工作而不阻塞主线程，但 IntentService 不能并行地处理多任务，只能依次处理，在完成所有的任务后自动调用 onDestroy()方法，无须调用

stopSelf()或 stopSelf(startId)方法。代码如下：

```
public class MyIntentService extends IntentService {
    public MyIntentService(String name) {
        super(name);
    }
    @Override
    protected void onHandleIntent(@Nullable Intent intent) {
        //doSomething
    }
}
```

在 MainActivity 中开启 Service 的代码如下：

```
Intent intent = new Intent(MainActivity.this,MyIntentService.class);
startService(intent);
```

11.4 开发实践：工厂通风系统界面切换

11.4.1 开发设计

本任务在工厂通风系统的主界面显示数据，同时在主界面中打开子界面，在子界面中采集数据后将数据传回给主界面。本任务利用 Android 的 Activity 之间的数据传递来实现该功能。工厂通风系统界面如图 11.3 所示，界面布局结构如图 11.4 所示。

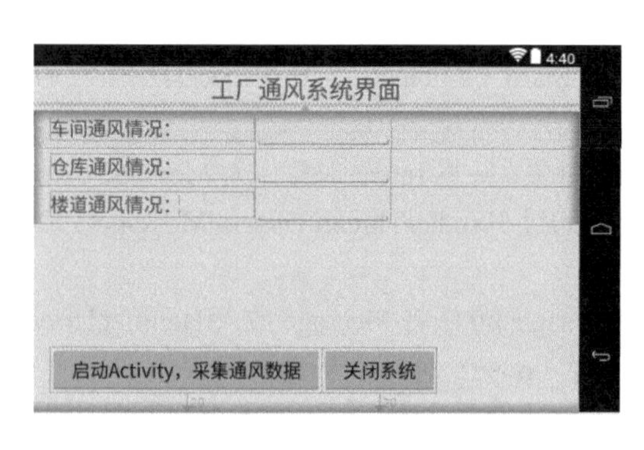

图 11.3 工厂通风系统界面

图 11.4 界面布局结构

本任务主要编写 MainActivity.java、SubActivity.java 程序文件，以及 activity_main.xml、content_main.xml 布局文件中，另外还需要修改 AndroidManifest.xml 项目配置文件。项目目录结构如图 11.5 所示。

项目开发步骤如下：

（1）编写 activity_main.xml 布局文件，设置界面显示的控件布局。

（2）编写 content_main.xm 布局文件，设置弹出子界面显示的控件布局。

（3）在 AndroidManifest.xml 项目配置文件中添加设置 SubActivity。

（4）编写 SubActivity.java 程序文件，设置“接收”按钮的 setOnClickListener 监听器，通过 setResult()方法向 MainActivity 返回文本编辑控件中设置的参数，设置“撤销”按钮通过 setResult()方法返回空。

（5）编写 MainActivity.java 程序文件，在 onCreate()方法中通过设置“启动 Activity，采集通风数据”按钮的 setOnClickListener 监听器，实例化一个 Intent，再通过 startActivityForResult()启动一个子界面。

（6）覆写 onActivityResult()方法，由 requestCode 参数确定不同的子界面，通过 data.getData()方法获取子界面传递的数据。

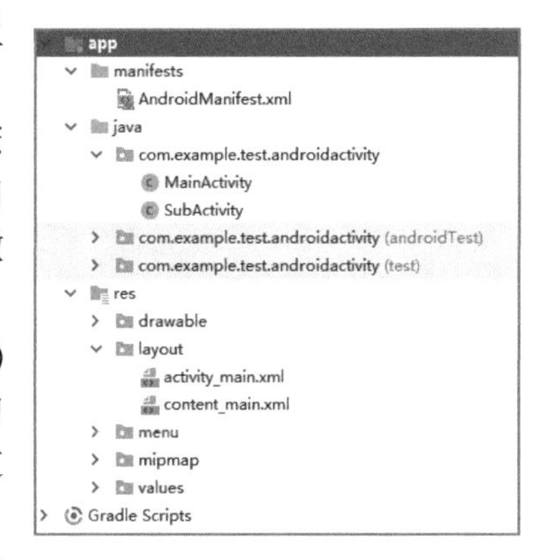

图 11.5　项目目录结构

11.4.2　功能实现

1. 布局设计

本任务有两个布局文件，分别是 activity_main.xml 和 content_main.xml，其中，主界面布局代码如下：

```
<?xml version="1.0" encoding="utf-8"?>
<RelativeLayout xmlns:android="http://schemas.android.com/apk/res/android"
    android:layout_width="match_parent"
    android:layout_height="match_parent"
    android:background="#e3e3e3">
    <TextView
        android:id="@+id/text_header"
        android:layout_width="fill_parent"
        android:layout_height="wrap_content"
        android:layout_marginBottom="5dp"
        android:layout_marginTop="5dp"
        android:gravity="center"
        android:text="@string/app_name"
        android:textSize="26sp" />
    <LinearLayout
        android:layout_width="match_parent"
        android:layout_height="match_parent"
        android:layout_below="@+id/text_header"
        android:orientation="vertical">
        <LinearLayout
            android:layout_width="match_parent"
            android:layout_height="wrap_content"
            android:orientation="horizontal">
```

```
        <TextView
            android:id="@+id/textView1"
            android:layout_width="150dp"
            android:layout_height="wrap_content"
            android:layout_marginLeft="20dp"
            android:text="车间通风情况："
            android:textSize="20sp" />
        <EditText
            android:id="@+id/editText1"
            android:layout_width="150dp"
            android:layout_height="wrap_content"
            android:editable="false"
            android:inputType="none"/>
    </LinearLayout>
    <LinearLayout
        android:layout_width="match_parent"
        android:layout_height="wrap_content"
        android:orientation="horizontal">
        <TextView
            android:id="@+id/textView2"
            android:layout_width="150dp"
            android:layout_height="wrap_content"
            android:layout_marginLeft="20dp"
            android:text="仓库通风情况："
            android:textSize="20sp" />
        <EditText
            android:id="@+id/editText2"
            android:layout_width="150dp"
            android:layout_height="wrap_content"
            android:editable="false"
            android:inputType="none"/>
    </LinearLayout>
    <LinearLayout
        android:layout_width="match_parent"
        android:layout_height="wrap_content"
        android:orientation="horizontal">
        <TextView
            android:id="@+id/textView3"
            android:layout_width="150dp"
            android:layout_height="wrap_content"
            android:layout_marginLeft="20dp"
            android:text="楼道通风情况："
            android:textSize="20sp" />
        <EditText
            android:id="@+id/editText3"
            android:layout_width="150dp"
            android:layout_height="wrap_content"
```

```
                android:editable="false"
                android:inputType="none" />
        </LinearLayout>
    </LinearLayout>
    <Button
        android:id="@+id/btn_start"
        android:layout_width="wrap_content"
        android:layout_height="wrap_content"
        android:layout_alignParentBottom="true"
        android:layout_marginLeft="20dp"
        android:padding="5dp"
        android:textSize="20sp"
        android:layout_marginBottom="20dp"
        android:text="    启动 Activity，采集通风数据    " />
    <Button
        android:id="@+id/btn_close"
        android:layout_width="wrap_content"
        android:layout_height="wrap_content"
        android:layout_alignParentBottom="true"
        android:layout_toRightOf="@+id/btn_start"
        android:padding="5dp"
        android:textSize="20sp"
        android:layout_marginBottom="20dp"
        android:text="    关闭系统    " />
</RelativeLayout>
```

2. 修改项目配置文件

```
<?xml version="1.0" encoding="utf-8"?>
<manifest xmlns:android="http://schemas.android.com/apk/res/android"
    package="com.example.test.androidactivity">
    <application
        android:allowBackup="true"
        android:icon="@mipmap/ic_launcher"
        android:label="@string/app_name"
        android:roundIcon="@mipmap/ic_launcher_round"
        android:supportsRtl="true"
        android:theme="@style/AppTheme">
        <activity
            android:name=".MainActivity"
            android:label="@string/app_name"
            android:theme="@style/AppTheme.NoActionBar">
            <intent-filter>
                <action android:name="android.intent.action.MAIN" />
                <category android:name="android.intent.category.LAUNCHER" />
            </intent-filter>
```

```
        </activity>
        <activity
            android:name ="SubActivity"
            android:theme="@style/AppTheme"
            android:screenOrientation="landscape"/>
    </application>
</manifest>
```

3．程序设计

（1）MainActivity.java 的代码如下：

```
public class MainActivity extends Activity {
    private static final int SUBACTIVITY1 = 1;
    private static final int SUBACTIVITY2 = 2;            //定义请求码（requestCode）
    EditText textView1;
    EditText textView2;
    EditText textView3;
    @Override
    public void onCreate(Bundle savedInstanceState) {
        super.onCreate(savedInstanceState);
        //去掉 TitleBar
        requestWindowFeature(Window.FEATURE_NO_TITLE);
        setContentView(R.layout.activity_main);
        textView1 =(EditText)findViewById(R.id.editText1);
        textView2 =(EditText)findViewById(R.id.editText2);
        textView3 =(EditText)findViewById(R.id.editText3);
        final Button btn1=(Button)findViewById(R.id.btn_start);
        btn1.setOnClickListener(new View.OnClickListener() {
            @Override
            public void onClick(View v) {
                Intent intent = new Intent(MainActivity.this,SubActivity.class);
                startActivityForResult(intent, SUBACTIVITY1);//以 sub-activity 启动子 Activity
            }
        });
    }
    @Override                                              //处理返回值
    protected void onActivityResult(int requestCode,int resultCode,Intent data){
        super.onActivityResult(requestCode, resultCode, data);
        switch (requestCode) {
            case SUBACTIVITY1:
                if(resultCode == RESULT_OK){
                    Uri uriData = data.getData();
                    textView1.setText(uriData.toString());
                    textView2.setText(uriData.toString());
                    textView3.setText(uriData.toString());
                }
```

```
                break;
            case SUBACTIVITY2:
                break;
            default:
                break;
        }
    }
}
```

（2）SubActivity.java 的代码如下：

```
public class SubActivity extends Activity{
    @Override
    public void onCreate(Bundle savedInstanceState) {
        super.onCreate(savedInstanceState);
        //去掉 TitleBar
        requestWindowFeature(Window.FEATURE_NO_TITLE);
        setContentView(R.layout.content_main);
        final EditText editText =(EditText)findViewById(R.id.edit);
        final Button btnOK=(Button)findViewById(R.id.ok);
        final Button btnCancel=(Button)findViewById(R.id.cancel);
        btnOK.setOnClickListener(new View.OnClickListener() {
            @Override
            public void onClick(View v) {
                String uriString = editText.getText().toString();
                Uri data = Uri.parse(uriString);                    //设置返回值
                Intent result = new Intent(null,data);
                setResult(RESULT_OK, result);
                finish();
            }
        });
        btnCancel.setOnClickListener(new View.OnClickListener() {
            @Override
            public void onClick(View v) {
                setResult(RESULT_CANCELED, null);
                finish();
            }
        });
    }
}
```

11.5　任务验证

在 Android Studio 开发环境中打开本项目的例程，编译通过后运行程序，程序运行效果如图 11.6 所示。单击“启动 Activity，采集通风数据”按钮，在弹出的界面输入通风数据，如图 11.7 所示。

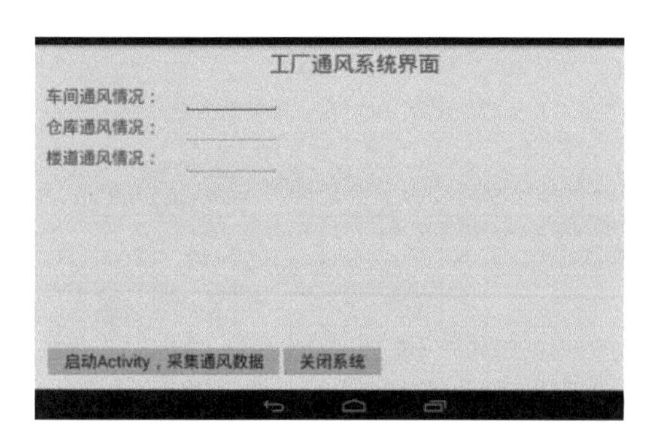

图 11.6　程序运行效果

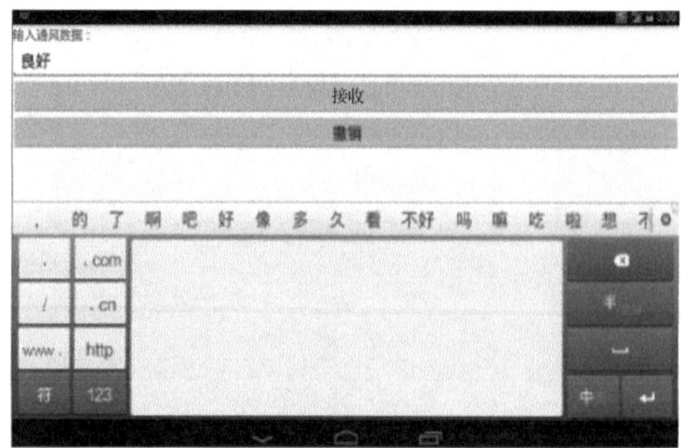

图 11.7　输入通风数据

单击“接收”按钮后，主界面会显示更新后的通风数据，如图 11.8 所示。

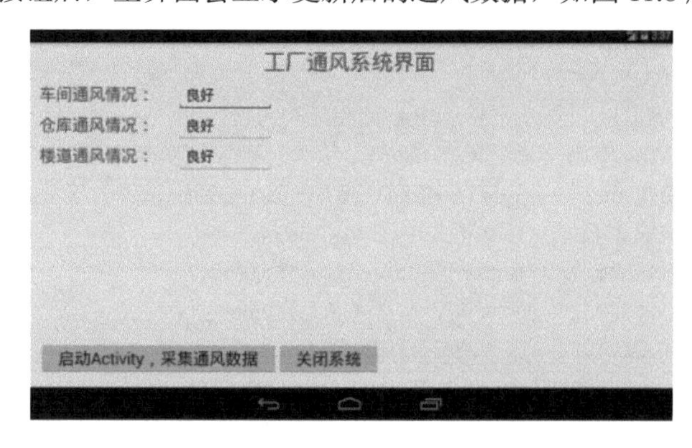

图 11.8　主界面显示更新后的通风数据

11.6　开发小结

本任务主要介绍了 Activity 的生命周期与创建、Activity 间的数据传递与交互、Service 与 Thread 的使用，以及 IntentService 的使用。通过本任务的学习，读者可以掌握 Activity 间的数据传递与交互方法，实现工厂通风系统界面的切换。

11.7　思考与拓展

（1）请谈谈 Android 中 Service 和 Thread 之间的区别。

（2）请尝试为界面中的“关闭系统”按钮增加功能，单击“关闭系统”按钮后退出主界面。

任务 12

工厂火警监测系统界面的设计

本任务主要介绍 Android 中的 Intent、BroadcastReceiver、ContentProvider 组件。通过本任务的学习，读者可掌握全局广播和本地广播、ContentProvider 的概念及流程，以及 Intent 的基本使用方法，实现工厂火警监测系统界面的设计。

12.1 开发场景：如何设计工厂火警监测系统的界面

在工厂、图书馆，酒店、博物馆和家庭住宅等场所，火警报警器已经是必不可少的装置。本任务实现的是一个简单的工厂火警监测系统界面的设计。

12.2 开发目标

（1）知识目标：熟悉 BroadcastReceiver、ContentProvider 的概念及流程，以及 Intent 的基本使用方法。

（2）技能目标：掌握 Intent 广播消息的方法。

（3）任务目标：熟悉 BroadcastReceiver、ContentProvider 的概念及流程，掌握 Intent 的基本使用方法，实现工厂火警监测系统界面的设计。

12.3 原理学习：Intent、BroadcastReceiver、ContentProvider 组件

12.3.1 Intent 组件

1. Intent 组件的机制

Android 中的 Intent（意图）组件可用于应用程序之间或应用程序内部的交互与通信。Intent 负责对应用程序中操作的动作、动作涉及数据、附加数据进行描述，Android 系统根据 Intent 的描述来找到对应的组件，将 Intent 传递给调用的组件。Intent 不仅可用于应用程序之间，也用于应用程序内部的 Activity 和 Service 之间的交互。Intent 起到中介的作用，

类似于消息、事件通知，充当Activity、Service、BroadcastReceiver之间通信的桥梁，提供组件互相调用的相关信息，实现调用者与被调用者之间的通信。Intent组件机制如图12.1所示。

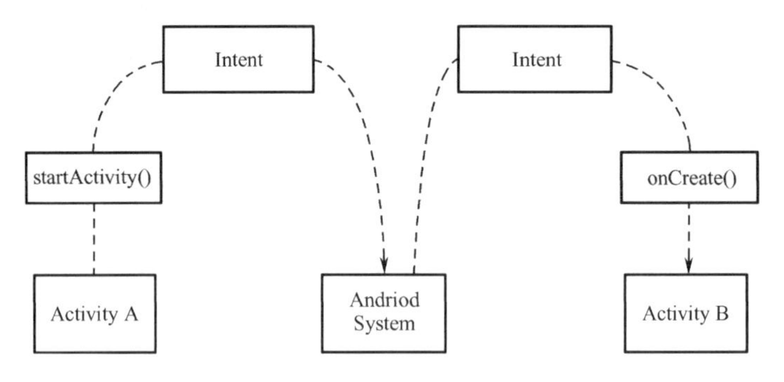

图12.1 Intent组件的机制

Intent可分为显式或隐式两类。显式的Intent指定了组件名字，由应用程序通过setComponent()方法或setClass()方法来指定具体的目标组件，在构造Intent对象时就指定了接收者。

```
Intent intent = new Intent(getContext(), SecondActivity.Class);
startActivity(intent);
```

隐式的Intent没指定明确的组件来处理Intent，需要通过比较Intent与应用组件中的Intent Filter来决定由哪个组件来处理Intent，Intent的发送者在构造Intent对象时，不知道、也不关心接收者是谁。例如：

```
Intent intent = new Intent();
intent.setAction(“test.intent.IntentTest”);
startActivity(intent);
```

目标组件（如Activity、Service、BroadcastReceiver）通过设置它们的Intent Filter来确定要处理的Intent。如果一个组件没有定义Intent Filter，那么它只能处理显式的Intent，只有定义了Intent Filter的组件才能处理隐式的Intent。

一个Intent包含的信息主要由以下6个部分组成：

（1）Action：要执行的动作。

（2）Data：动作要操作的数据。

（3）Category：动作的附加信息。

（4）Extras：其他所有附加信息的集合。

（5）Type：Intent的数据类型。

（6）Component：指定的目标组件的名称，如要执行的动作、类别、数据、附加信息等。

2. Intent Filter

应用程序的组件通过声明一个或多个Intent Filter（意图过滤器）来告诉Android自己能响应、处理哪些隐式的Intent，每个Intent Filter描述该组件可以接收什么类型的Intent。

Intent解析机制主要是通过查找在AndroidManifest.xml声明的Intent Filter以及其中定义的Intent，最终找到匹配的Intent。Android主要是通过Intent的Action、Type、Category这三

个属性来解析 Intent 的，方法如下：

（1）如果 Intent 指明了 Action，则目标组件 Intent Filter 的 Action 属性中就必须包含这个 Action，否则不能匹配。

（2）如果 Intent 没有提供 Type，系统将从 Data 中得到 Type。和 Action 一样，目标组件的 Type 属性中必须包含 Intent 的 Type，否则不能匹配。

（3）如果 Intent 指定了一个或多个 Category，这些类别必须全部出现在目标组件的 Category 属性中。例如，Intent 中包含了两个 Category，如 LAUNCHER_CATEGORY 和 ALTERNATIVE_CATEGORY，则目标组件的 Category 属性中必须至少包含这两个 Category。

一个 Intent 对象只能指定一个 Action，而目标组件中的一个 Intent Filter 可以指定多个 Action，目标组件中的 Action 属性不能为空，否则它将阻止所有的 Intent。

一个 Intent 对象的 Action 必须和目标组件中的 Intent Filter 中的某一个 Action 匹配，才能匹配。如果 Intent Filter 的 Action 属性为空，则无法匹配；如果 Intent 对象不指定 Action，并且 Intent Filter 的 Action 属性不为空，则可以匹配。

3. Intent 的方法

Intent 可以启动一个 Activity，也可以启动、停止或绑定一个 Service，还可以发起一个 Broadcast（广播），Intent 的方法如表 12.1 所示。

表 12.1　Intent 的方法

针对的组件	相应的方法
Activity	startActivity()、startActivityForResult()
Service	startService()、stopService()、bindService()
Broadcast	sendBroadcast ()、sendOrderBroadcast()

4. Intent 的属性

Intent 的属性包括 Action（动作）、Data（数据）、Category（分类）、Type（类型）、Component（组件）、Extras（附加信息），其中最常用的属性是 Action 和 Data。

（1）Action 属性。对于有如下声明的 Detail Activity：

```
<activity android:name=".DetailActivity">
    <intent-filter>
        <action android:name="com.xxx.intent.action.DETAIL"/>
        <category android:name="android.intent.category.DEFAULT"/>
    </intent-filter>
</activity>
```

该 DetailActivity 在其<intent-filter>中声明 action，即目标组件 action。如果需要做一个跳转的动作，就需要在 Intent 中指定目标组件的 action，代码如下：

```
public void gotoDetailActivity(View view) {
    Intent intent = new Intent("com.xxx.intent.action.DETAIL");
    startActivity(intent);
}
```

在为Intent指定相应的Action，然后调用startActivity()方法后，系统会根据Action跳转到对应的Activity。除了自定义的Action，Intent也内含了很多默认的Action，例如：

```
public static final String ACTION_VIEW = "android.intent.action.VIEW";
public static final String ACTION_PICK = "android.intent.action.PICK";
public static final String ACTION_ANSWER = "android.intent.action.ANSWER";
public static final String ACTION_CALL = "android.intent.action.CALL";
```

（2）Data属性。Data表示动作要操作的数据。例如，与浏览器交互的代码如下：

```
public void invokeWebBrowser(View view) {
    Intent intent = new Intent(Intent.ACTION_VIEW);
    intent.setData(Uri.parse("http://www.qq.com"));
startActivity(intent);
}
public void invokeWebSearch(View view) {
    Intent intent = new Intent(Intent.ACTION_WEB_SEARCH);
    intent.putExtra(SearchManager.QUERY, "android");    //搜索关键字
startActivity(intent);
}
```

上面两种方法分别启动浏览器后打开指定网页，以及搜索关键字，对应的Action分别是Intent.ACTION_VIEW和Intent.ACTION_WEB_SEARCH，前者需指定相应的网页地址，后者需指定关键字。对于搜索关键字来说，浏览器会按照自己默认的搜索引擎进行搜索。

注意，在打开网页时，为Intent指定一个Data属性，其实就是指定要操作的数据，采用的是URI形式，可以将一个指定前缀的字符串转换成特定的URI形式，如“tel:”表示电话号码类。例如：

```
public void call(View view) {
    Intent intent = new Intent(Intent.ACTION_CALL);
    intent.setData(Uri.parse("tel:12345678"));
    startActivity(intent);
}
```

5. Intent用法示例

下述示例代码有些功能需要在项目配置文件中申请相关权限。

（1）调用拨号程序。代码如下：

```
//拨打10086电话
Uri testUri = Uri.parse("tel:10086");
Intent mIntent = new Intent(Intent.ACTION_DIAL, testUri);
startActivity(mIntent );
```

（2）发送短信或彩信。代码如下：

```
//向10086发送内容为“测试信息”的短信
Uri testUri = Uri.parse("smsto:10086");
```

```
Intent mIntent = new Intent(Intent.ACTION_SENDTO, testUri);
mIntent.putExtra("sms_body", "测试信息");
startActivity(mIntent);
```

（3）通过浏览器打开网页。代码如下：

```
//打开百度主页
Uri testUri = Uri.parse("http://www.baidu.com");
Intent mIntent = new Intent(Intent.ACTION_VIEW, testUri);
startActivity(mIntent);
```

（4）播放多媒体。代码如下：

```
Intent mIntent = new Intent(Intent.ACTION_VIEW);
Uri testUri = Uri.parse("file:///sdcard/test.mp3");
mIntent.setDataAndType(testUri , "audio/mp3");
startActivity(mIntent);
```

（5）拍照。代码如下：

```
//打开拍照程序
Intent mIntent= new Intent(MediaStore.ACTION_IMAGE_CAPTURE);
startActivityForResult(mIntent, 0);
```

12.3.2　BroadcastReceiver 组件

1. BroadcastReceiver 的工作原理

Android 中的 Broadcast 是一种用于在应用程序之间广播消息的组件；BroadcastReceiver 是接收并响应广播消息的组件，对广播的消息进行过滤接收并响应，它不包含任何用户界面，可以通过启动 Activity 或者 Notification 通知用户接收重要消息。在 Notification 中有多种方法提示用户，如背景灯闪动、设备振动、发出声音或在状态栏上放置一个持久的图标。BroadcastReceiver 组件如图 12.2 所示。

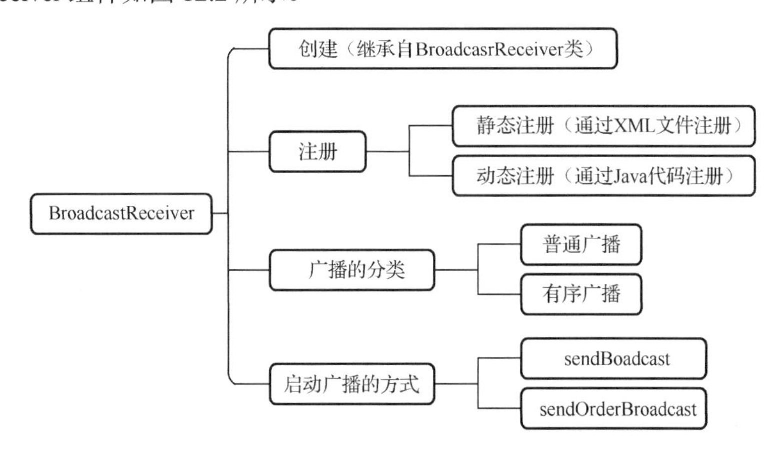

图 12.2　BroadcastReceiver 组件

在广播消息时，把要广播的消息和用于过滤的属性（如 Action、Category）装入一个 Intent

对象中，然后调用 Context.sendBroadcast()、sendOrderBroadcast()或 sendStickyBroadcast()方法把 Intent 对象广播出去。

当广播 Intent 后，所有已经注册的 BroadcastReceiver 都会检查 Intent Filter 是否与发送的 Intent 相匹配，若匹配则调用 BroadcastReceiver 的 onReceive()方法来接收 Intent 对象，因此在定义 BroadcastReceiver 时，通常需要实现 onReceive()方法。

2. LocalBroadcastReceiver

BroadcastReceiver 是针对应用程序间、应用程序与系统间、应用程序内部进行通信的一种方式，LocalBroadcastReceiver 仅在自己的应用程序内广播消息，只有自己的应用程序才能收到。

LocalBroadcastReceiver 不能静态注册，只能动态注册，在发送广播消息和注册 Local Broadcast Receiver 时，分别调用 LocalBroadcastReceiver 的 sendBroadcast()方法和 register Receiver()方法。

3. BroadcastReceiver 的定义

在定义 BroadcastReceiver 时，需要覆写 BroadcastReceiver 类的 onReceive()方法。BroadcastReceiver 接收到广播的消息后，会自动调用 onReceive()方法。onReceive()方法一般要与其他组件交互，如 Notification、Service 等。BroadcastReceiver 运行在 UI 线程，因此 onReceive()方法不能执行耗时的操作，否则将导致 ANR（Application Not Responding）问题。定义 BroadcastReceiver 的代码如下：

```
public class mBroadcastReceiver extends BroadcastReceiver {
    //覆写 onReceive()方法。接收到广播的消息后，则自动调用该方法
    @Override
    public void onReceive(Context context, Intent intent) {
        //接收到广播的消息后的操作
    }
}
```

4. BroadcastReceiver 的注册方式

BroadcastReceiver 的方式注册有以下两种：

（1）静态注册：在 AndroidManifest.xml 中用<receiver>标签声明注册，并在<intent-filter>标签内设置 Intent Filter。

（2）动态注册：先定义并设置一个 Intent Filter 对象，在注册时调用 Context.registerReceiver()方法，在取消时调用 Context.unregisterReceiver()方法。

不论采用静态注册的方式还是采用动态注册的方式来注册 BroadcastReceiver，在程序退出时都要注销 BroadcastReceiver，否则在下次启动程序时可能会有多个 BroadcastReceiver。另外，若在使用 sendBroadcast()方法时指定了接收权限，则只有在 AndroidManifest.xml 中用<user-permission>标签声明拥有此权限的 BroadcastReceiver 才能接收到广播的消息。

同样，若在注册 BroadcastReceiver 时指定了接收权限，则只有在 AndroidManifest.xml 中用<user-permission>标签声明拥有此权限的 Intent 对象所广播的消息才能被这个 BroadcastReceiver 所接收。

静态注册是指在 AndroidManifest.xml 中通过<receive>标签声明 BroadcastReceiver，其属性说明代码如下：

```
<receiver
    //此 BroadcastReceiver 能否接收其他 App 广播的消息
    android:enabled=["true" | "false"]
    //默认值是由 BroadcastReceiver 中有无 Intent Filter 决定的：如果有 Intent Filter，则默认值为 true；否则为 false
    android:exported=["true" | "false"]
    android:icon="drawable resource"
    android:label="string resource"
    //继承自 BroadcastReceiver 类
    android:name=".mBroadcastReceiver"
    //具有相应权限的 Intent 对象广播的消息才能被此 BroadcastReceiver 接收
    android:permission="string"
    //BroadcastReceiver 运行所处的进程，默认为应用程序的进程，也可以指定独立的进程
    android:process="string" >
    //用于指定此 BroadcastReceiver 可以接收的广播消息类型，如设置为网络状态改变
    <intent-filter>
        <action android:name="android.net.conn.CONNECTIVITY_CHANGE" />
    </intent-filter>
</receiver>
```

静态注册示例代码如下：

```
<receiver
    //定义 mBroadcastReceiver
    android:name=".mBroadcastReceiver" >
    //用于接收网络状态改变时广播的消息
    <intent-filter>
        <action android:name="android.net.conn.CONNECTIVITY_CHANGE" />
    </intent-filter>
</receiver>
```

动态注册是指调用 registerReceiver()方法来注册 BroadcastReceiver，代码如下：

```
//在 onResume()中注册 BroadcastReceiver
@Override
    protected void onResume(){
        super.onResume();
        //实例化 BroadcastReceiver 与 Intent Filter
        mBroadcastReceiver mBroadcastReceiver = new mBroadcastReceiver();
        IntentFilter intentFilter = new IntentFilter();
        //设置接收广播消息的类型
        intentFilter.addAction(android.net.conn.CONNECTIVITY_CHANGE);
        //动态注册：调用 registerReceiver()方法
        registerReceiver(mBroadcastReceiver, intentFilter);
    }
    //注销 BroadcastReceiver
```

```
    @Override
    protected void onPause() {
        super.onPause();
        //注销在 onResume()方法中注册的 BroadcastReceiver
        unregisterReceiver(mBroadcastReceiver);
    }
}
```

BroadcastReceiver 最好通过 Activity 的 onResume()方法注册，通过 onPause()方法注销，有注册就必然要有注销，否则会导致内存泄漏，也不允许重复注册、重复注销。

12.3.3 ContentProvider 组件

ContentProvider 是 Android 提供的一种标准的共享数据的机制。在 Android 中，每一个应用程序的资源都是私有的，但应用程序可以通过 ContentProvider 来访问其他应用程序的私有数据（私有数据可以存储在文件系统的文件中，也可以存放在 SQLite 数据库中）。应用程序、ContentResolver 和 ContentProvider 的关系如图 12.3 所示。

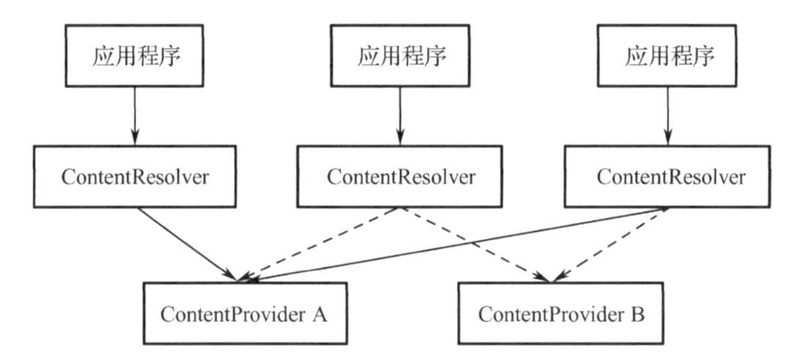

图 12.3 应用程序、ContentResolver 与 ContentProvider 的关系

使用 ContentProvider 的方式有以下两种：

（1）通过 ContentResolver 使用 ContentProvider。

（2）通过 Context.getContentResolver 使用 ContentProvider。

Android 内部也提供了一些内置的 ContentProvider，能够为应用程序提供重要的数据。使用 ContentProvider 对外共享数据的好处是可以统一数据的访问方式。

进程间共享数据的操作主要有插入、删除、获取、修改（更新）数据，因此 ContentProvider 的核心方法有 4 个，如下所示：

```
//4 个核心方法
//向 ContentProvider 插入数据
public Uri insert(Uri uri, ContentValues values)
//删除 ContentProvider 中的数据
public int delete(Uri uri, String selection, String[] selectionArgs)
//更新 ContentProvider 中的数据
public int update(Uri uri, ContentValues values, String selection, String[] selectionArgs)
//获取 ContentProvider 中的数据
public Cursor query(Uri uri, String[] projection, String selection, String[] selectionArgs,    String sortOrder)
```

```
//注：(1) 上述 4 个方法由外部进程调用，运行在 ContentProvider 进程的 Binder 线程池中。
//(2) 在多线程并发访问时，需要实现线程同步。若 ContentProvider 采用 SQLite 数据库存储数据，则不需要同步，SQLite 数据库已经实现了线程同步。如果有多个 SQLite 数据库，则需要实现线程同步，因为 SQLite 数据库之间无法进行线程同步。若 ContentProvider 的数据存储在内存中，则需要实现线程同步。
//其他的两个方法
//创建 ContentProvider 后或打开系统后其他进程第一次访问 ContentProvider 时，由系统调用运行在
//ContentProvider 进程中的主线程，不能做耗时操作
public boolean onCreate()
//得到数据类型，返回当前 URL 所代表数据的类型
public String getType(Uri exam)
```

12.4　开发实践：工厂火警监测系统界面的设计

12.4.1　开发设计

本任务假设已经在监测区域安装了多个火警报警器，工厂火警监测系统通过 4 个按钮（Button）模拟火警报警器。工厂火警监测系统的主界面的如图 12.4 所示，主界面的布局结构如图 12.5 所示。

图 12.4　工厂火警监测系统的主界面

图 12.5　工厂火警监测系统主界面的布局结构

本任务主要编写 MainActivity.java、MyBroadcastReceiver.java 程序文件，以及 activity_main.xml 布局文件，还需要修改 AndroidManifest.xml 项目配置文件。项目目录结构如图 12.6 所示。

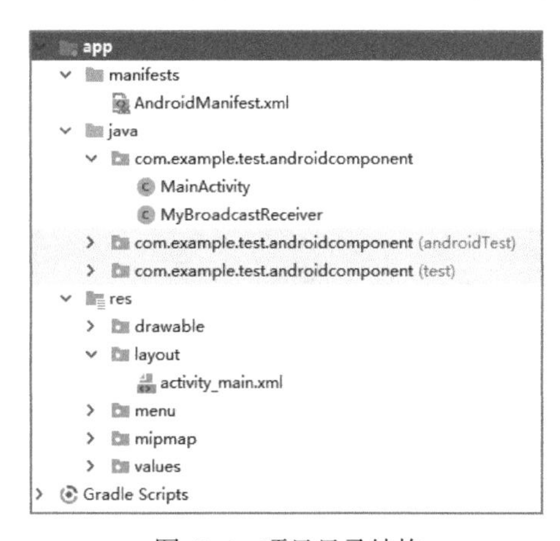

图 12.6　项目目录结构

项目开发步骤如下：

（1）编写 activity_main.xml 布局文件，设置界面显示的控件布局。

（2）编写 MyBroadcastReceiver.java 程序，定义一个 BroadcastReceiver，覆写 onReceive() 方法，在该方法中通过 intent.getStringExtra 获取

广播的消息，并调用 Toast 显示消息内容。

（3）在 AndroidManifest.xml 项目配置文件中注册并设置 BroadcastReceiver。

（4）编写 MainActivity.java 程序，设置四个按钮的 setOnClickListener 监听器，实例化 Intent，识别广播消息的字符串标识，在 intent.putExtra 中设置要广播的消息，通过 sendBroadcast(intent) 方法广播消息。

12.4.2 功能实现

当发生火灾时，即按钮被按下，通过广播消息来通知火警监测点，然后通过屏幕弹框模拟火灾响应。

1. 界面布局设计

界面布局的代码如下：

```
<?xml version="1.0" encoding="utf-8"?>
<RelativeLayout xmlns:android="http://schemas.android.com/apk/res/android"
    android:layout_width="match_parent"
    android:layout_height="match_parent"
    android:background="#e3e3e3">
    <TextView
        android:id="@+id/text_header"
        android:layout_width="fill_parent"
        android:layout_height="wrap_content"
        android:layout_marginBottom="5dp"
        android:layout_marginTop="5dp"
        android:gravity="center"
        android:text="@string/app_name"
        android:textSize="26sp" />
    <LinearLayout
        android:layout_width="match_parent"
        android:layout_height="match_parent"
        android:layout_below="@+id/text_header"
        android:layout_marginLeft="30dp"
        android:orientation="vertical">
        <LinearLayout
            android:layout_width="match_parent"
            android:layout_height="wrap_content"
            android:orientation="horizontal">
            <ImageView
                android:id="@+id/imageView1"
                android:layout_width="120dp"
                android:layout_height="60dp"
                android:scaleType="fitXY"
                android:src="@drawable/p1" />
            <Button
```

```
                android:id="@+id/btn1"
                android:layout_marginLeft="5dp"
                android:layout_width="100dp"
                android:layout_height="wrap_content"
                android:padding="3dp"
                android:text="车间着火" />
        </LinearLayout>
    <LinearLayout
        android:layout_width="match_parent"
        android:layout_height="wrap_content"
        android:orientation="horizontal">
        <ImageView
            android:id="@+id/imageView2"
            android:layout_marginTop="2dp"
            android:layout_width="120dp"
            android:layout_height="60dp"
            android:scaleType="fitXY"
            android:src="@drawable/p2" />
        <Button
            android:id="@+id/btn2"
            android:layout_marginLeft="5dp"
            android:layout_width="100dp"
            android:layout_height="wrap_content"
            android:padding="3dp"
            android:text="仓库着火" /></LinearLayout>
    <LinearLayout
        android:layout_width="match_parent"
        android:layout_height="wrap_content"
        android:orientation="horizontal">
        <ImageView
            android:id="@+id/imageView3"
            android:layout_marginTop="2dp"
            android:layout_width="120dp"
            android:layout_height="60dp"
            android:scaleType="fitXY"
            android:src="@drawable/p3" />
        <Button
            android:id="@+id/btn3"
            android:layout_marginLeft="5dp"
            android:layout_width="100dp"
            android:layout_height="wrap_content"
            android:padding="3dp"
            android:text="楼道着火" /></LinearLayout>
    <LinearLayout
        android:layout_width="match_parent"
        android:layout_height="wrap_content"
        android:orientation="horizontal">
```

```
            <ImageView
                android:id="@+id/imageView4"
                android:layout_marginTop="2dp"
                android:layout_width="120dp"
                android:layout_height="60dp"
                android:scaleType="fitXY"
                android:src="@drawable/p4" />
            <Button
                android:id="@+id/btn4"
                android:layout_marginLeft="5dp"
                android:layout_width="100dp"
                android:layout_height="wrap_content"
                android:padding="3dp"
                android:text="食堂着火" /></LinearLayout>
        </LinearLayout>
</RelativeLayout>
```

2. 清单文件

清单文件除了定义应用程序的 Application 和 Activity，还定义了 MyBroadcastReceiver，代码如下：

```
<?xml version="1.0" encoding="utf-8"?>
<manifest xmlns:android="http://schemas.android.com/apk/res/android"
    package="com.x210.broadcast"
    android:versionCode="1"
    android:versionName="1.0">
    <uses-sdk android:minSdkVersion="8" />
    <application
        android:icon="@drawable/ic_launcher"
        android:label="@string/app_name">
        <activity
            android:theme="@style/AppTheme"
            android:screenOrientation="landscape"
            android:name=".BroadcastReceiverDemo"
            android:label="@string/app_name">
            <intent-filter>
                <action android:name="android.intent.action.MAIN" />
                <category android:name="android.intent.category.LAUNCHER" />
            </intent-filter>
        </activity>
        <!-- receiver 节点 -->
        <receiver android:name=".MyBroadcastReceiver">
            <intent-filter>
                <!-- Intent 过滤的动作 -->
                <action android:name="com.x210.broadcast" />
            </intent-filter>
        </receiver>
```

```
    </application>
</manifest>
```

3. 程序设计

（1）MainActivity.java 的代码如下：

```
public class MainActivity.java extends Activity {
    static final String ReceiverAction = "com.x210.broadcast";
    private Button button1;
    private Button button2;
    private Button button3;
    private Button button4;
    /*Called when the activity is first created.*/
    @Override
    public void onCreate(Bundle savedInstanceState) {
        super.onCreate(savedInstanceState);
        //去掉 TitleBar
        requestWindowFeature(Window.FEATURE_NO_TITLE);
        setContentView(R.layout.main);
        button1 = (Button)findViewById(R.id.btn1);
        button1.setOnClickListener(new View.OnClickListener() {
            @Override
            public void onClick(View v) {
                Intent intent = new Intent (ReceiverAction);//识别广播消息的字符串标识
                intent.putExtra("message", "车间发出火警……");
                sendBroadcast(intent);//发送广播信息
            }
        });
        button2 = (Button)findViewById(R.id.btn2);
        button2.setOnClickListener(new View.OnClickListener() {
            @Override
            public void onClick(View v) {
                Intent intent = new Intent (ReceiverAction);//识别广播消息的字符串标识
                intent.putExtra("message", "仓库发出火警……");
                sendBroadcast(intent);//发送广播信息
            }
        });
        button3 = (Button)findViewById(R.id.btn3);
        button3.setOnClickListener(new View.OnClickListener() {
            @Override
            public void onClick(View v) {
                Intent intent = new Intent (ReceiverAction);//识别广播消息的字符串标识
                intent.putExtra("message", "楼道发出火警……");
                sendBroadcast(intent);//发送广播信息
            }
        });
        button4 = (Button)findViewById(R.id.btn4);
```

```
            button4.setOnClickListener(new View.OnClickListener() {
                @Override
                public void onClick(View v) {
                    Intent intent = new Intent (ReceiverAction);//识别广播消息的字符串标识
                    intent.putExtra("message", "食堂发出火警……");
                    sendBroadcast(intent);//发送广播信息
                }
            });
        }
    }
```

（2）MyBroadcastReceiver.java 的代码如下：

```
public class MyBroadcastReceiver extends BroadcastReceiver {
    @Override
    public void onReceive(Context context, Intent intent) {
        String msg = intent.getStringExtra("message");
        Toast.makeText(context, "Receiver 收到："+msg, Toast.LENGTH_LONG).show();
    }
}
```

12.5 任务验证

在 Android Studio 开发环境中打开本任务的例程，编译通过后运行程序。程序运行效果如图 12.7 所示。

图 12.7　程序运行效果

12.6 开发小结

本任务主要介绍 Android 的 Intent、BroadcastReceiver、ContentProvider 等组件，实现了工厂火警监测系统界面的设计。

12.7　思考与拓展

（1）BroadcastReceiver 有几种注册方式？

（2）请尝试在本任务例程的基础上，当 BroadcastReceiver 接收到火警报警后，再发出一条疏散工人的消息。

任务13

设备列表管理界面的设计

本任务介绍 Fragment 的一些基本概念以及用法，帮助读者熟悉并掌握基于 Fragment 架构的界面设计方法。在本任务的界面设计中，首先设计一个主页，然后通过 Fragment 对每个 Tab 项进行布局，在不同的 Tab 项之间进行切换，从而实现设备列表管理界面的设计。

13.1 开发场景：如何使用 Fragment 设计界面

Fragment 是为了适应大屏幕的平板电脑出现的，在界面设计中，可以把 Fragment 看成一个小型的 Activity（又称为 Activity 片段）。Fragment 可以把屏幕划分成几块，然后分组进行模块化管理，从而可以更加方便地在运行过程中动态地更新 Activity。

13.2 开发目标

（1）知识目标：熟悉 Fragment 的基本概念、生命周期及使用方法。
（2）技能目标：理解 Fragment 的生命周期，掌握 Fragment 的使用方法。
（3）任务目标：掌握基于 Fragment 的界面设计方式，实现设备列表管理界面的设计。

13.3 原理学习：基于 Fragment 的界面设计

13.3.1 Fragment 的基本概念

Fragment 是 Activity 片段，可以在运行过程中修改 Activity 的外观，并在由 Activity 管理的返回（Back）栈中保留这些更改，Fragment 的优势是其布局可以在不同设备上的适配。Fragment 应用界面如图 13.1 所示。

从图 13.1 中可以看到：在平板电脑中，只有一个 Activity（即 Activity A），包含了两个 Fragment，分别是 Fragment A 和 Fragment B；在智能手机中就需要两个 Activity，分别是 Activity A（包含 Fragment A）和 Activity B（包含 Fragment B）。每个 Fragment 都有自己的一套生命

周期方法，并处理各自的用户输入事件，因此在平板电脑中使用一个 Activity 就可以了，左侧是条目列表，右边是条目详情。

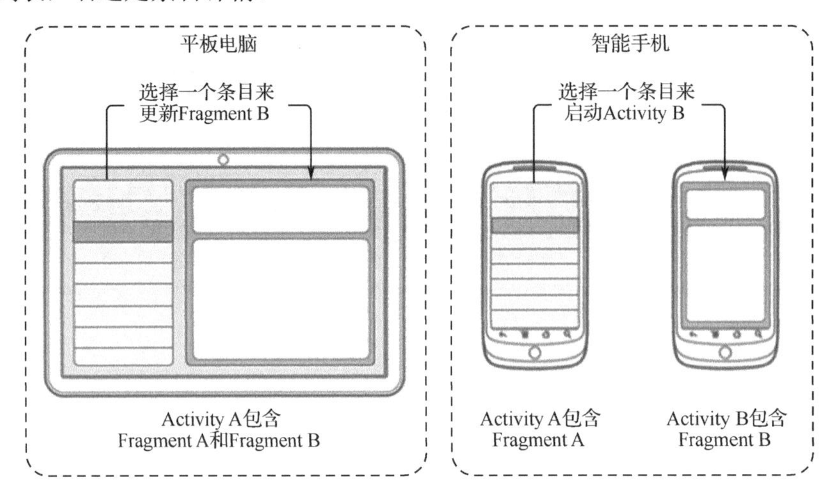

图 13.1 Fragment 应用界面

开发者可以在一个 Activity 中使用多个 Fragment 来构建多窗格的 UI，也可以在多个 Activity 中重复使用某个 Fragment。Fragment 有自己的生命周期，能接收自己的输入，开发者可以在 Activity 运行时添加或删除 Fragment。

13.3.2 Fragment 的生命周期

前文介绍了 Activity 的生命周期，由于 Fragment 托管在 Activity 中，所以两者的生命周期有很多相似的地方，例如都有 onCreate()、onStart()、onPause()、onDestroy()等方法。因为 Fragment 托管在 Activity 中，所以 Fragment 多了两个方法，即 onAttach()和 onDetach()。Fragment 生命周期的常用方法如表 13.1 所示。

表 13.1 Fragment 生命周期的常用方法

方 法	说 明
onAttach()	Fragment 和 Activity 建立关联时调用该方法，用于将 Fragment 托管到 Activity 中
onCreate()	系统会在创建 Fragment 时调用该方法，用于初始化资源文件等
onCreateView()	系统会在 Fragment 首次绘制用户界面时调用该方法。要想为 Fragment 绘制 UI，则该方法返回的 View 必须是 Fragment 布局的根视图
onViewCreated()	在 Fragment 被绘制后调用该方法，用于初始化控件资源
onActivityCreated()	执行完 onCreate()、onCreateView()和 onViewCreated()方法后，也就是 Activity 被渲染绘制后调用该方法
onPause()	系统在暂停 Fragment 时调用该方法，可在该方法内确认在当前用户会话结束后仍然有效的更改
onDestroyView()	当 Fragment 中的布局被移除时调用该方法
onDetach()	当 Fragment 和 Activity 解除关联时调用该方法

除了 onCreateView()方法，如果覆写了其他的方法，则必须调用父类实现相应的方法。Fragment 的生命周期如图 13.2 所示。

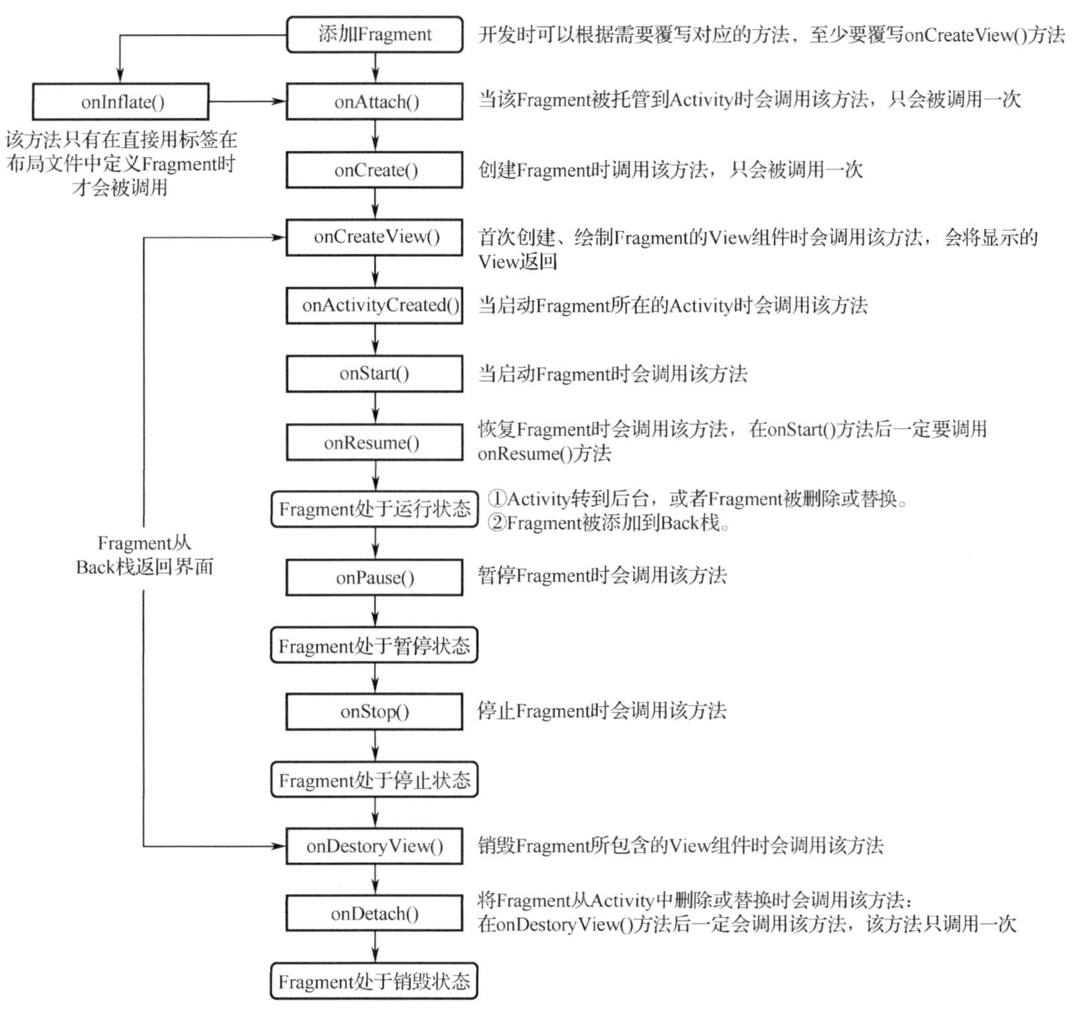

图 13.2　Fragment 的生命周期

13.3.3　Fragment 的使用方式

Fragment 既可以嵌套到 Activity 中，也可以嵌套到另外一个 Fragment 中，但这个被嵌套的 Fragment 需要嵌套到 Activity 中，即 Fragment 还是最终要嵌套到 Activity 中。Fragment 受 Activity 的生命周期影响，但也有自己的生命周期。Fragment 的用法有两种：静态用法和动态用法。

1．静态用法

静态用法的流程如图 13.3 所示。

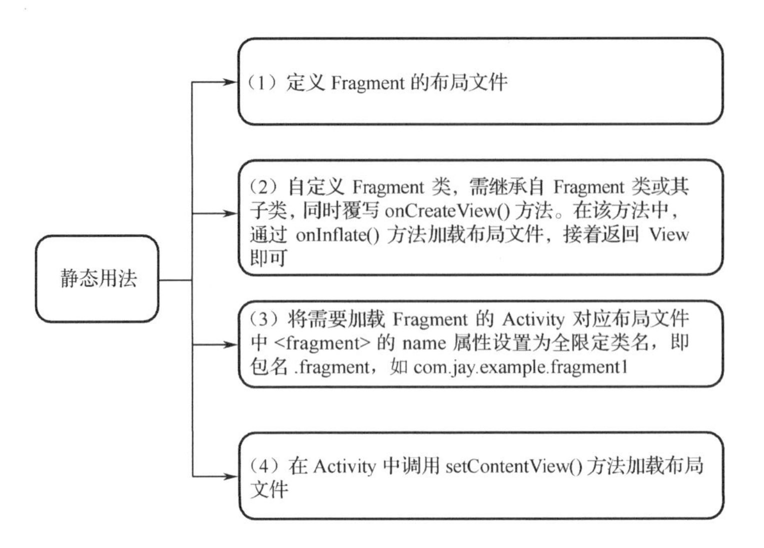

图 13.3　静态用法的流程

布局文件 fragment1.xml 的代码如下：

```
<LinearLayout
    android:layout_height="match_parent"
    android:layout_weight="match_parent"
    android:background="#FDD835"
    android:gravity="center">
    <TextView
        android:layout_width="wrap_content"
        android:layout_height="wrap_content"
        android:text="第一个 Fragment"
        android:textColor="#5E35B1"
        android:textSize="25sp" />
</LinearLayout>
```

上述代码只有一个 LinearLayout，并加入了一个 TextView。这里再新建布局文件 fragment2.xml，代码如下：

```
<LinearLayout
        android:layout_height="match_parent"
        android:layout_weight="match_parent"
        android:background="#3949AB"
        android:gravity="center" >
        <TextView
            android:layout_width="wrap_content"
            android:layout_height="wrap_content"
            android:text="第二个 Fragment"
            android:textColor="#C0CA33"
            android:textSize="25sp" />
    </LinearLayout>
```

新建一个类 Fragment1，这个类继承自 Fragment 类，代码如下：

```
public class Fragment1 extends Fragment {
    @Override
    public View onCreateView(LayoutInflater inflater, ViewGroup container, Bundle savedInstanceState) {
        return inflater.inflate(R.layout.fragment1, container, false);
    }
}
```

可以看到，在 onCreateView()方法中加载了 fragment1.xml 的布局文件，布局文件 fragment2.xml 的用法也是一样的。新建一个类 Fragment2，代码如下：

```
public class Fragment2 extends Fragment {
    @Override
    public View onCreateView(LayoutInflater inflater, ViewGroup container, Bundle savedInstanceState) {
        return inflater.inflate(R.layout.fragment2, container, false);
    }
}
```

打开或新建 activity_main.xml 并将其作为主 Activity 的布局文件，在里面加入类 Fragment1 和类 Fragment2 的引用，使用 android:name 前缀来引用 Fragment1 和 Fragment2。

```
<LinearLayout xmlns:android="http://schemas.android.com/apk/res/android"
    android:layout_width="match_parent"
    android:layout_height="match_parent"
    android:orientation="horizontal"
    android:baselineAligned="false" >
    <fragment
        android:id="@+id/fragment1"
        android:name="com.example.demo.Fragment1"
        android:layout_width="0dp"
        android:layout_height="match_parent"
        android:layout_weight="1" />
    <fragment
        android:id="@+id/fragment2"
        android:name="com.example.demo.Fragment2"
        android:layout_width="0dp"
        android:layout_height="match_parent"
        android:layout_weight="1" />
</LinearLayout>
```

新建 MainActivity 并将其作为程序的主 Activity，MainActivity 中的代码是自动生成的，如下所示：

```
public class MainActivity extends Activity {
    @Override
    protected void onCreate(Bundle savedInstanceState) {
        super.onCreate(savedInstanceState);
        setContentView(R.layout.activity_main);
    }
}
```

现在运行程序就会看到，一个 Activity 包含了两个 Fragment，这两个 Fragment 平分了整个屏幕，如图 13.4 所示。

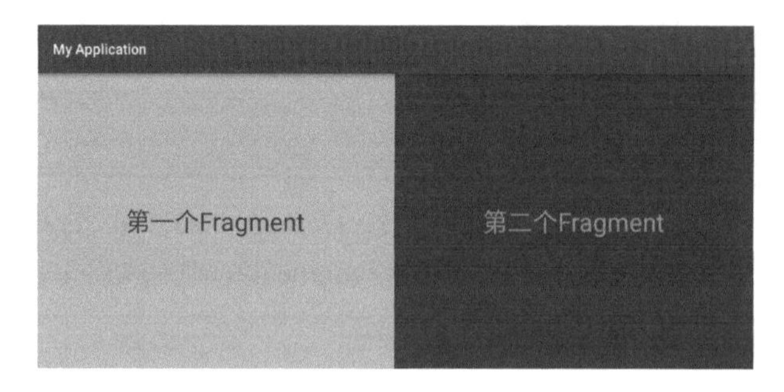

图 13.4　两个 Fragment 平分整个屏幕的效果

2. 动态用法

使用 Fragment 的动态用法时，不需要在 Fragment 的 XML 布局文件中添加对 Fragment 的引用。动态用法在静态用法代码的基础上进行修改，将 Fragment 的引用删除，只保留最外层的 LinearLayout，并为它添加一个 ID。删除后代码如下：

```
<LinearLayout xmlns:android="http://schemas.android.com/apk/res/android"
    android:id="@+id/main_layout"
    android:layout_width="match_parent"
    android:layout_height="match_parent"   />
```

打开 MainActivity，修改其中的代码，如下所示：

```
public class MainActivity extends Activity {
    @Override
    protected void onCreate(Bundle savedInstanceState) {
        super.onCreate(savedInstanceState);
        setContentView(R.layout.activity_main);
        Display display = getWindowManager().getDefaultDisplay();
        if (display.getWidth() >= display.getHeight()) {
            Fragment1 fragment1 = new Fragment1();
            getFragmentManager().beginTransaction().replace(R.id.main_layout, fragment1).commit();
        } else {
            Fragment2 fragment2 = new Fragment2();
            getFragmentManager().beginTransaction().replace(R.id.main_layout, fragment2).commit();
        }
    }
}
```

上述代码获取了屏幕的宽度和高度，如果屏幕宽度大于或等于高度就添加 Fragment1，否则添加 Fragment2。

动态加载 Fragment 的流程如图 13.5 所示。

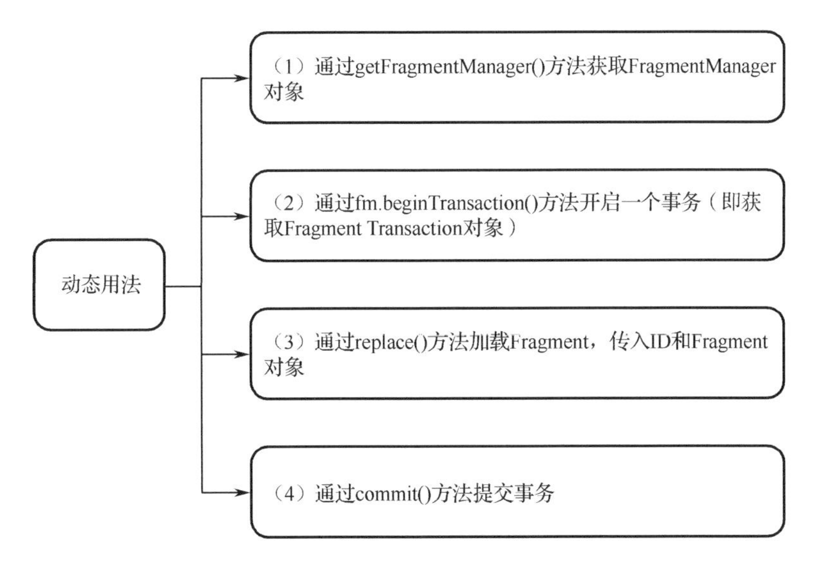

图 13.5　动态加载 Fragment 的流程

运行效果如图 13.6 所示。

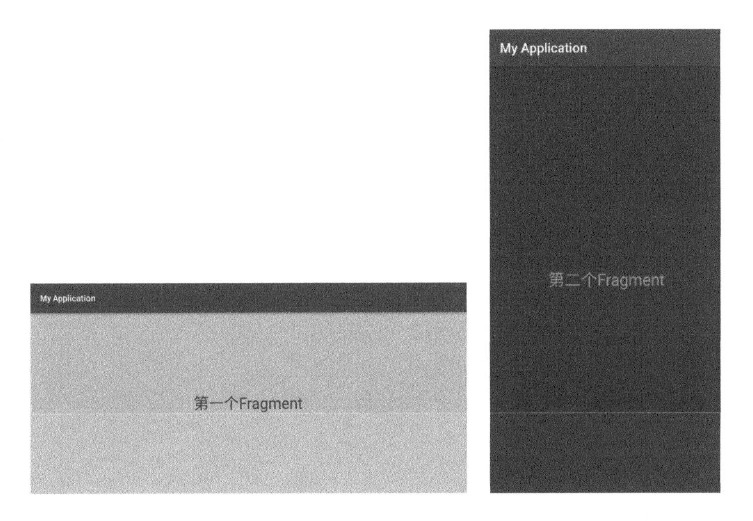

图 13.6　Fragment 动态加载效果

Android 是通过 FragmentManager 来管理 Fragment 的，FragmentManager 的使用方法如下：

（1）通过 findFragmentById()方法（在 Activity 布局中提供 UI 的 Fragment 时使用）或 findFragmentByTag()方法（在 Activity 布局中提供或不提供 UI 的 Fragment 时使用）获取 Activity 中的 Fragment。

（2）通过 popBackStack()从 Back 栈中弹出 Fragment。

（3）通过 addOnBackStackChangedListener()方法注册一个监听器，用于监听 Back 栈的变化：也可以通过 FragmentManager 打开一个 FragmentTransaction 来执行某些操作，如添加和删除 Fragment。

13.3.4 Fragment通信

1. 组件获取

Fragment可以在多个Activity内使用，但Fragment会直接绑定到包含它的Activity上。Fragment可以通过getActivity()方法访问Activity，在Activity布局中实现查找视图等任务。例如：

```
View listView = getActivity().findViewById(R.id.list);
```

Activity可以通过findFragmentById()方法或findFragmentByTag()方法从FragmentManager调用Fragment中的方法。例如：

```
ExampleFragment fragment = (ExampleFragment) getFragmentManager().findFragmentById(R.id.example_
fragment);
```

2. 事件回调

当Fragment需要和Activity共享事件时，可以在Fragment内定义一个回调接口，并要求绑定Fragment的Activity实现该回调接口，Activity可以根据需要通过该回调接口与Fragment共享事件。

例如，如果一个应用的Activity中有两个Fragment，一个用于显示文章列表（FragmentArticleList），另一个用于显示文章内容（FragmentArticleDetail）。FragmentArticleList必须在文章列表中的某选项被选中后告知Activity，以便Activity告知FragmentArticleDetail显示该文章内容。在本例中，OnArticleClickedListener回调接口是在FragmentArticleList内声明的，代码如下：

```
public static class FragmentArticleList extends ListFragment {
    public interface OnArticleClickedListener {
        public void onArticleClicked(Uri articleUri);
    }
}
```

在绑定的Activity中实现OnArticleClickedListener回调接口并覆写onArticleClicked()方法，将来自FragmentArticleList的事件通知给FragmentArticleDetail。为确保Activity实现回调接口，FragmentArticleList的onAttach()方法（系统在向Activity添加Fragment时调用的方法）会通过传递到onAttach()方法中的Activity来实例化OnArticleClickedListener，代码如下：

```
public static class FragmentArticleList extends ListFragment {
OnArticleClickedListener mListener;
    @Override
    public void onAttach(Activity activity) {
        super.onAttach(activity);
        try {
            mListener = (OnArticleClickedListener) activity;
        } catch (ClassCastException e) {
```

```
                Log.e(TAG, e.toString());
            }
        }
    }
```

如果 Activity 未实现回调接口，则 Fragment 会抛出 ClassCastException。在实现回调接口后，mListener 成员会保留对 Activity 的 OnArticleClickedListener 的引用，以便 FragmentArticleList 通过调用 OnArticleClickedListener 回调接口定义的方法与 Activity 共享事件。如果 FragmentArticleList 是 ListFragment 的一个扩展，则用户在每次选中文章列表中的某个选项时，系统都会调用 Fragment 中的 onListItemClick()方法，该方法会调用 onArticleClicked()方法和 Activity 共享事件。代码如下：

```
public static class FragmentArticleList extends ListFragment {
    OnArticleClickedListener mListener;
    @Override
    public void onListItemClick(ListView l, View v, int position, long id) {
        Uri noteUri = ContentUris.withAppendedId(ArticleColumns.CONTENT_URI, id);
        mListener.onArticleClicked(noteUri);
    }
}
```

13.4 开发实践：设备列表管理界面设计

13.4.1 开发设计

本任务界面的中间区域是一个布局容器，采用 FrameLayout，然后通过 replace()方法或 add()方法将一个 Fragment 添加到布局容器。该 Fragment 中有一个 ListView，当单击这个 ListView 中的某一项时，布局容器中的 Fragment 就会通过 replace()方法替换成包含详细信息的 Fragment。如果仅通过 replace()方法，则不会保存第一个 Fragment 的状态。可以通过 Fragment 中的 addToBackStack()方法和 popBackStack()方法来解决这个问题，可以将被替换的 Fragment 添加到 Fragment 栈中；当返回时，可调用 popBackStack()方法来弹出 Fragment 栈。项目运行界面如图 13.7 所示。

图 13.7 项目运行界面

本任务需要编写的文件较多，如表 13.2 所示。

表 13.2　需要编写的文件说明表

文 件 名	功 能 说 明	文 件 名	功 能 说 明
Data.java	数据类文件	MainActivity.java	主界面文件
MyAdapter.java	数据适配器文件	NewContentFragment.java	内容详情片断（NewContentFragment）类文件
NewListFragment.java	实现设备列表片断（NewListFragment）类文件	activity_main.xml	主界面布局文件
fg_context.xml	内容详情布局文件	fg_devlist.xml	设备列表布局文件
list_item.xml	列表标题布局文件	—	—

项目目录结构如图 13.8 所示。

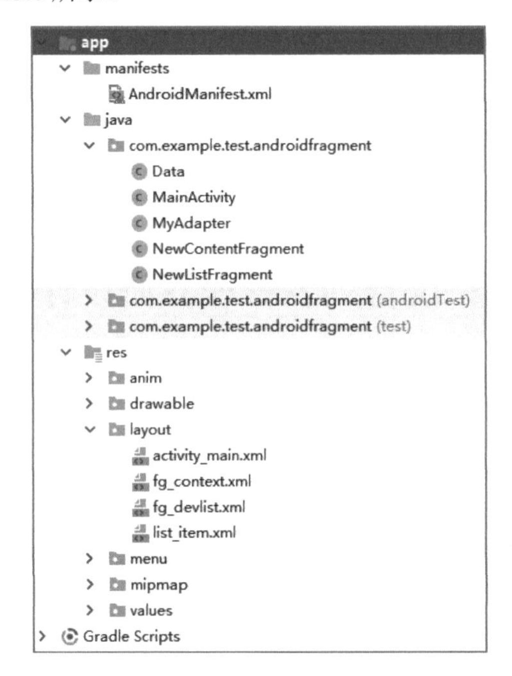

图 13.8　项目目录结构

项目开发步骤如下：

（1）编写 fg_context.xml 内容详情布局文件，设置内容详情片段（NewContentFragment）界面显示的控件布局。

（2）编写 fg_devlist.xml 设备列表布局文件，设置设备列表片段（NewListFragment）界面显示的控件布局。

（3）编写 activity_main.xml 主界面布局文件，设置主界面布局。

（4）编写 Data.java 程序文件代码，定义一个数据类，用于处理 NewListFragment 中每项数据的标题与内容。

（5）编写 MyAdapter.java 文件代码，用于实现设备列表的数据适配器。

（6）编写 NewListFragment.java 代码，通过 onCreateViewInflater.inflate 加载布局文件，

设置 NewListFragment 的 setOnItemClickListener 监听器。

（7）能过 onItemClick()方法中的 fManager.beginTransaction()方法开启一个事务，通过该事务的 replace()方法向布局容器加入 NewContentFragment，通过 commit()方法提交事务。

（8）编写 NewContentFragment.java 代码：通过 inflater.inflate 加载布局文件，通过 getArgument()方法获取传递过来的 Bundle 对象。

（9）编写 MainActivity.java 程序文件代码，通过 onCreate()方法中的 getFragmentManager()方法获取 FragmentManager，通过 bindViews()方法设置 FrameLayout。

（10）实例化一个 NewListFragment，通过 fManager.beginTransaction()方法开启一个事务，通过事务的 replace()方法向布局容器加入 NewListFragment，通过 commit()方法提交事务。

13.4.2 功能实现

1. 项目布局设计

（1）Fragment 和 Activity 的布局设计。内容详情布局文件 fg_context.xml 的代码如下：

```
<?xml version="1.0" encoding="utf-8"?>
<LinearLayout xmlns:android="http://schemas.android.com/apk/res/android"
    android:layout_width="match_parent"
    android:layout_height="match_parent"
    android:orientation="vertical">
    <ImageView
        android:id="@+id/air_image1"
        android:layout_width="150dp"
        android:layout_height="150dp"
        android:layout_gravity="center"
        android:background="@drawable/cam"/>
    <TextView
        android:id="@+id/txt_content"
        android:layout_width="match_parent"
        android:layout_height="match_parent"
        android:gravity="center"
        android:textColor="@color/blue"
        android:textSize="20sp" />
</LinearLayout>
```

（2）设备列表布局文件 fg_devlist.xml 的代码如下：

```
<?xml version="1.0" encoding="utf-8"?>
<LincarLayout xmlns:android="http://schemas.android.com/apk/res/android"
    android:layout_width="match_parent"
    android:layout_height="match_parent"
    android:background="@color/white"
    android:orientation="horizontal">
    <ListView
        android:id="@+id/list_news"
```

```
        android:layout_width="match_parent"
        android:layout_height="match_parent"
        android:gravity="center"/>
</LinearLayout>
```

（3）主界面文件 MainActivity.java 的代码如下：

```
<RelativeLayout xmlns:android="http://schemas.android.com/apk/res/android"
    xmlns:tools="http://schemas.android.com/tools"
    android:layout_width="match_parent"
    android:layout_height="match_parent"
    tools:context=".MainActivity">
    <TextView
        android:id="@+id/txt_title"
        android:layout_width="match_parent"
        android:layout_height="56dp"
        android:background="@color/blue"
        android:textColor="@color/white"
        android:text="系统设备列表"
        android:textSize="20sp"
        android:textStyle="bold"
        android:gravity="center"/>
    <FrameLayout
        android:id="@+id/fl_content"
        android:layout_width="match_parent"
        android:layout_height="match_parent"
        android:layout_below="@id/txt_title"/>
</RelativeLayout>
```

2．Java 项目设计

（1）数据类文件 Data.java 的代码如下：

```
package com.example.test.androidfragment;
public class Data {
    private String new_title;
    private String new_content;
    public Data(){}
    public Data(String new_title, String new_content) {
        this.new_title = new_title;
        this.new_content = new_content;
    }
    public String getNew_title() {
        return new_title;
    }
    public String getNew_content() {
        return new_content;
    }
    public void setNew_title(String new_title) {
```

```
            this.new_title = new_title;
        }
        public void setNew_content(String new_content) {
            this.new_content = new_content;
        }
}
```

（2）数据适配器文件 MyAdapter.java 的代码如下：

```
package com.example.test.androidfragment;
import android.content.Context;
import android.view.LayoutInflater;
import android.view.View;
import android.view.ViewGroup;
import android.widget.BaseAdapter;
import android.widget.TextView;
import java.util.List;
public class MyAdapter extends BaseAdapter{
        private List<Data> mData;
        private Context mContext;
        public MyAdapter(List<Data> mData, Context mContext) {
            this.mData = mData;
            this.mContext = mContext;
        }
        @Override
        public int getCount() {
            return mData.size();
        }
        @Override
        public Object getItem(int position) {
            return null;
        }
        @Override
        public long getItemId(int position) {
            return position;
        }
        @Override
        public View getView(int position, View convertView, ViewGroup parent) {
            ViewHolder viewHolder;
            if(convertView == null){
             convertView = LayoutInflater.from(mContext).inflate(R.layout.list_item,parent,false);
                viewHolder = new ViewHolder();
                viewHolder.txt_item_title = (TextView) convertView.findViewById(R.id.txt_item_title);
                convertView.setTag(viewHolder);
            }else{
                viewHolder = (ViewHolder) convertView.getTag();
            }
            viewHolder.txt_item_title.setText(mData.get(position).getNew_title());
```

```
            return convertView;
        }
        private class ViewHolder{
            TextView txt_item_title;
        }
}
```

（3）设备列表片断（NewListFragment）类文件的代码如下：

```
package com.example.test.androidfragment;
import android.app.Fragment;
import android.app.FragmentManager;
import android.app.FragmentTransaction;
import android.os.Bundle;
import android.view.LayoutInflater;
import android.view.View;
import android.view.ViewGroup;
import android.widget.AdapterView;
import android.widget.ListView;
import android.widget.TextView;
import java.util.ArrayList;
public class NewListFragment extends Fragment implements AdapterView.OnItemClickListener {
    private FragmentManager fManager;
    private ArrayList<Data> datas;
    private ListView list_news;
    public NewListFragment(FragmentManager fManager, ArrayList<Data> datas) {
        this.fManager = fManager;
        this.datas = datas;
    }
    @Override
    public View onCreateView(LayoutInflater inflater, ViewGroup container, Bundle savedInstanceState) {
        View view = inflater.inflate(R.layout.fg_devlist, container, false);
        list_news = (ListView) view.findViewById(R.id.list_news);
        MyAdapter myAdapter = new MyAdapter(datas, getActivity());
        list_news.setAdapter(myAdapter);
        list_news.setOnItemClickListener(this);
        return view;
    }
    @Override
    public void onItemClick(AdapterView<?> parent, View view, int position, long id) {
        FragmentTransaction fTransaction = fManager.beginTransaction();
        NewContentFragment ncFragment = new NewContentFragment();
        Bundle bd = new Bundle();
        bd.putString("content", datas.get(position).getNew_content());
        ncFragment.setArguments(bd);
        //获取 Activity 的控件
        TextView txt_title = (TextView) getActivity().findViewById(R.id.txt_title);
        txt_title.setText(datas.get(position).getNew_title());
```

```
            //加上 Fragment 替换动画
            fTransaction.setCustomAnimations(R.anim.fragment_slide_left_enter,
R.anim.fragment_slide_left_exit);
            fTransaction.replace(R.id.fl_content, ncFragment);
            //调用 addToBackStack 将 Fragment 添加到栈中
            fTransaction.addToBackStack(null);
            fTransaction.commit();
        }
}
```

（4）设备详情页片断（NewContentFragment）类文件的代码如下：

```
package com.example.test.androidfragment;
import android.app.Fragment;
import android.os.Bundle;
import android.view.LayoutInflater;
import android.view.View;
import android.view.ViewGroup;
import android.widget.TextView;
public class NewContentFragment extends Fragment {
    @Override
    public View onCreateView(LayoutInflater inflater, ViewGroup container, Bundle savedInstanceState) {
        View view = inflater.inflate(R.layout.fg_context, container, false);
        TextView txt_content = (TextView) view.findViewById(R.id.txt_content);
        //通过 getArgument()方法获取传递过来的 Bundle 对象
        txt_content.setText(getArguments().getString("content"));
        return view;
    }
}
```

（5）主界面文件 MainActivity.java 的代码如下：

```
package com.example.test.androidfragment;
import android.app.FragmentManager;
import android.app.FragmentTransaction;
import android.content.Context;
import android.os.Bundle;
import android.support.v7.app.AppCompatActivity;
import android.widget.FrameLayout;
import android.widget.TextView;
import android.widget.Toast;
import java.util.ArrayList;
public class MainActivity extends AppCompatActivity {
    private TextView txt_title;
    private FrameLayout fl_content;
    private Context mContext;
    private ArrayList<Data> datas = null;
    private FragmentManager fManager = null;
```

```
    private long exitTime = 0;
    @Override
    protected void onCreate(Bundle savedInstanceState) {
        super.onCreate(savedInstanceState);
        setContentView(R.layout.activity_main);
        mContext = MainActivity.this;
        fManager = getFragmentManager();
        bindViews();
        datas = new ArrayList<Data>();
        for (int i = 1; i <= 10; i++) {
            Data data = new Data("监控摄像机" + i, i + "设备 IP:191.11.1.3");
            datas.add(data);
        }
        NewListFragment nlFragment = new NewListFragment(fManager, datas);
        FragmentTransaction ft = fManager.beginTransaction();
        ft.replace(R.id.fl_content, nlFragment);
        ft.commit();
    }
    private void bindViews() {
        txt_title = (TextView) findViewById(R.id.txt_title);
        fl_content = (FrameLayout) findViewById(R.id.fl_content);
    }
    //返回处理：判断 Fragment 栈中是否有 Fragment。若没有 Fragment 则双击退出程序，像 Toast 提
示一样；若有 Fragment 则通过 popBackStack()方法将 Fragment 弹出 Fragment 栈
    @Override
    public void onBackPressed() {
        if (fManager.getBackStackEntryCount() == 0) {
            if ((System.currentTimeMillis() - exitTime) > 2000) {
                Toast.makeText(getApplicationContext(), "再按一次退出程序",
                        Toast.LENGTH_SHORT).show();
                exitTime = System.currentTimeMillis();
            } else {
                super.onBackPressed();
            }
        } else {
            fManager.popBackStack();
            txt_title.setText("设备列表");
        }
    }
}
```

13.5 任务验证

在 Android Studio 开发环境中打开本任务的例程，编译通过后运行程序。系统设备列表如图 13.9 所示。

图 13.9　系统设备列表

单击其中一个条目，可切换到基于 Fragment 设计的内容详情界面，如图 13.10 所示。

图 13.10　内容详情界面

13.6　开发小结

本任务主要介绍了 Fragment 的基本概念、生命周期和使用方法。通过本任务的学习，读者可以熟悉并掌握基于 Fragment 的界面设计方法，从而实现设备列表管理界面的设计。

13.7　思考与拓展

（1）使用 Fragment 完全替换 Activity 真的比常规开发模式更好吗？

（2）Fragment 是什么？有什么缺点？简述它的生命周期。

（3）如何使用 Fragment？

任务 14

智能电表日志的记录

本任务主要介绍 Android 中的 SharedPreferences 和文件存储的使用方法。通过本任务的学习，读者可以熟悉并掌握 SharedPreferences 和文件存储的使用方法，实现智能电表日志的记录。

14.1 开发场景：如何实现智能电表日志的记录

能耗的计量、监测与管理是实现节能减排的基础。能耗管理系统可以通过互联网来对各类能耗进行精细计量、实时监测、智能处理和动态管控。例如，通过智能电表日志来分析能耗。如何记录智能电表日志？

14.2 开发目标

（1）知识目标：了解 SharedPreferences 和文件存储的使用方法。
（2）技能目标：掌握 SharedPreferences 和文件存储的使用方法。
（3）任务目标：实现智能电表日志的记录。

14.3 原理学习：SharedPreferences 及文件存储的使用

14.3.1 SharedPreferences

1. SharedPreferences 概述

SharedPreferences 是 Android 中的一个轻量级的存储类，是最简单的一种数据存储方式，常用来保存应用程序中的一些常用配置。例如，可以存储 Activity 的状态，在 Activity 处于暂停状态时，可以将此状态保存到 SharedPreferences 中；在重载 Activity 时，Android 系统可以通过 onSaveInstanceState()方法从 SharedPreferences 中将此状态取出。

SharedPreferences 可以通过 Android 系统生成一个 XML 文件，并将该 XML 文件保存到"/data/data/包名/shared_prefs"目录下。SharedPreferences 提供了常规数据类型（如 int、long、boolean、string 等）的保存接口。

SharedPreferences 的使用步骤如下：

（1）得到 SharedPreferences 对象。可通过以下三种方法来得到简单存储对象：

方法 1：

```
Context.getSharedPreferences(文件名称，操作模式);
```

在方法 1 中，如果文件名称不存在就会创建一个，操作模式有两种：

① MODE_PRIVATE：直接在把第二个参数写成 0 即可使用这种操作模式，这种操作模式表示只有当前的应用程序才可以对当前的文件进行读写。

② MODE_MULTI_PRIVATE：用于多个进程共同操作 SharedPreferences 的文件。

方法 2：

```
Activity.getPreferences(操作模式);
```

方法 2 可以将当前活动的类名作为 SharedPreferences 的文件名，通过"Activity.getSharedPreferences(String name, int mode);"可以来调用方法 2，并可传入自定义的名字。

方法 3：

```
PreferenceManager.getDefaultSharedPreferences(Context);
```

方法 3 可以使用当前程序的包名作为前缀来命名 SharedPreferences 文件。

（2）调用 SharedPreferences 对象的 edit()方法来获取一个 SharedPreferences.Editor 对象，代码如下：

```
SharedPreferences.Editor editor = getSharedPreferences("Person",Context.MODE_PRIVATE).edit();
```

（3）向 SharedPreferences.Editor 对象中添加数据，代码如下：

```
editor.putString("name", "Jack");                //以键值对的形式存储字符串
editor.putInt("age", 25);                        //以键值对的形式存储整数
editor.putBoolean("married", true);              //以键值对的形式存储布尔值
```

（4）调用 commit 方法可以提交添加的数据，代码如下：

```
editor.commit();
```

2．SharedPreferences 的使用方法

（1）存储数据。代码如下：

```
SharedPreferences sp = getSharedPreferences("person", Context.MODE_PRIVATE);
sp.edit().putString("name", "Jack").putInt("age", 25).commit();
```

也可以使用下面的方法来存储数据，代码如下：

```
SharedPreferences sp = getSharedPreferences("person", Context.MODE_PRIVATE);
Editor editor = sp.edit();
editor.putString("name", "Jack");
```

```
editor.putInt("age", 25);
editor.commit();
```

（2）读取数据。代码如下：

```
SharedPreferences sp = getSharedPreferences("person", Context.MODE_PRIVATE);
String name = sp.getString("name", null);
int age = sp.getInt("age", 0);
```

3．SharedPreferences 示例

界面布局的代码如下：

```
<?xml version="1.0" encoding="utf-8"?>
<LinearLayout xmlns:android="http://schemas.android.com/apk/res/android"
    android:layout_width="match_parent"
    android:layout_height="match_parent"
    android:orientation="vertical"
    android:padding="20dp">
    <EditText
        android:id="@+id/edt_name"
        android:layout_width="match_parent"
        android:layout_height="wrap_content"
        android:hint="请输入姓名" />
    <EditText
        android:id="@+id/edt_age"
        android:layout_width="match_parent"
        android:layout_height="wrap_content"
        android:hint="请输入年龄"
        android:inputType="number" />
    <LinearLayout
        android:layout_width="match_parent"
        android:layout_height="wrap_content"
        android:orientation="horizontal">
        <Button
            android:id="@+id/btn_save"
            android:layout_width="wrap_content"
            android:layout_height="wrap_content"
            android:text="保存参数" />
        <Button
            android:id="@+id/btn_read"
            android:layout_width="wrap_content"
            android:layout_height="wrap_content"
            android:text="读取参数" />
    </LinearLayout>
</LinearLayout>
```

SharedPreferences 界面如图 14.1 所示。

用户在姓名和年龄（使用 EditText 控件实现）对应的文本框中输入相应的值，单击“保存参数”按钮后相应的值就会保存到 SharedPreferences 中。当两个 EditText 中没有值时，单击“读取参数”按钮可以从 SharedPreferences 文件中读取姓名和年龄的值并显示在相应的控件上。

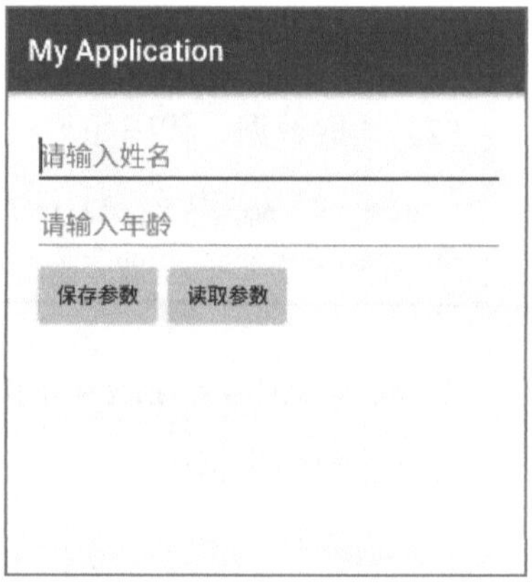

图 14.1　SharedPreferences 界面

MainActivity 的代码如下：

```
public class MainActivity extends AppCompatActivity {
    private EditText edtName;
    private EditText edtAge;
    private Button btnSave;
    private Button btnRead;
    @Override
    protected void onCreate(Bundle savedInstanceState) {
        super.onCreate(savedInstanceState);
        setContentView(R.layout.activity_main);
        edtName = findViewById(R.id.edt_name);
        edtAge = findViewById(R.id.edt_age);
        btnRead = findViewById(R.id.btn_read);
        btnSave = findViewById(R.id.btn_save);
        btnSave.setOnClickListener(new View.OnClickListener() {
            @Override
            public void onClick(View v) {
                String name = edtName.getText().toString();
                int age = Integer.parseInt(edtAge.getText().toString());
                SharedPreferences sp = getSharedPreferences("person", MODE_PRIVATE);
                SharedPreferences.Editor edit = sp.edit();
                edit.putString("name",name);
                edit.putInt("age", age);
                edit.commit();
            }
        });
        btnRead.setOnClickListener(new View.OnClickListener() {
            @Override
            public void onClick(View v) {
                SharedPreferences sp = getSharedPreferences("person", MODE_PRIVATE);
                String name = sp.getString("name","");
                int age = sp.getInt("age", 0);
                edtName.setText(name);
                edtAge.setText(String.valueOf(age));
            }
        });
    }
}
```

生成的 test.xml 的代码如下：

```
<?xml version='1.0' encoding='utf-8' standalone='yes' ?>
<map>
<string name="name">zhangbo</string>
<int name="age" value="24" />
</map>
```

14.3.2　文件存储

1．内部存储

如果将文件存储在内部存储中，则文件默认只能被创建该文件的应用程序访问，且一个应用程序所创建的所有文件都存储在和应用包名相同的目录下。也就是说，由应用程序创建的并存储在内部存储的文件，是与这个应用程序关联在一起的。内部存储一般用 Context 来获取和操作。访问内部存储的方法如下：

方法 1：用于获取内部存储的根目录，代码如下：

```
Environment.getDataDirectory() = /data;
```

方法 2：用于获取某个应用程序在内部存储中的“files”目录，代码如下：

```
getFilesDir().getAbsolutePath() = /data/user/0/packname/files;
```

方法 3：用于获取某个应用程序在内部存储中的“cache”目录，代码如下：

```
getCacheDir().getAbsolutePath() = /data/user/0/packname/cache;
```

方法 4：用于获取某个应用程序在内部存储中的自定义目录，代码如下：

```
getDir(“myFile”, MODE_PRIVATE).getAbsolutePath() = /data/user/0/packname/app_myFile;
```

内部存储的常用操作如下：

（1）列出所有已创建的文件，代码如下：

```
String[] files = Context.fileList();
for(String file : files) {
    Log.e(TAG, "file is "+ file);
}
```

（2）删除文件，代码如下：

```
if(Context.deleteFile(filename)) {
    Log.e(TAG, "delete file "+ filename + " sucessfully“);
} else {
    Log.e(TAG, "failed to delete file " + filename);
}
```

（3）创建一个目录，代码如下：

```
File workDir = Context.getDir(dirName, Context.MODE_PRIVATE);
Log.e(TAG, "workdir "+ workDir.getAbsolutePath();
```

2. 外部存储

外部存储可以通过物理媒介（如 SD 卡）实现，也可以由一部分内部存储封装而成，Android 设备可以有多个外部存储实例。

所有的 Android 设备都有外部存储和内部存储，很多中高端 Android 设备的自身存储已达到了 512 GB 以上，但在概念上仍然分成了内部存储和外部存储两部分，其实都在手机内部。不论 Android 设备是否有 SD 卡，总有外部存储和内部存储。Android 是通过相同的 API 来访问 SD 卡或者 Android 设备自身的外部存储的。访问外部存储的方法如下：

方法 1：获取外部存储的根目录。代码如下：

```
Environment.getExternalStorageDirectory().getAbsolutePath() = /storage/emulated/0
```

方法 2：获取外部存储的公共目录根目录，代码如下：

```
Environment.getExternalStoragePublicDirectory("").getAbsolutePath() = /storage/emulated/0
```

方法 3：获取某个应用程序在外部存储中的“files”目录，代码如下：

```
getExternalFilesDir("").getAbsolutePath() = /storage/emulated/0/Android/data/packname/files
```

方法 4：获取某个应用程序在外部存储中的“cache”目录，代码如下：

```
getExternalCacheDir().getAbsolutePath() = /storage/emulated/0/Android/data/packname/cache
```

注意：对于不同的外部设备，使用方法 4 获取的“cache”目录略有不同。

外部存储的文件可以被用户或者其他应用程序修改。外部存储的文件有两种类型：

（1）公共文件（Public Files）：公共文件可以被自由访问，当卸载应用程序之后，创建的公共文件仍然保留。

（2）私有文件（Private Files）：私有文件也能被其他应用程序访问，只不过一个应用程序的私有文件对其他应用程序来说是没有访问价值的。

3. 资源文件

Android 的资源文件主要包括字符串、颜色、数组、动画、布局、图像、图标、音频、视频，以及其他应用程序使用的组件。在 Android 开发中，资源文件的使用频率很高，资源文件所在的目录一般有下面三个：

（1）/res/drawable：存储图像和图标资源文件。

（2）/res/layout：存储用户布局资源文件，如 Widget。

（3）/res/values：存储字符串、颜色等资源文件。

下面介绍几种常用的资源文件及其使用方法。

（1）字符串。字符串存储在“/res/values/strings.xml”文件中，其格式比较简单，读取字符串的代码如下：

```
String str = getResources().getString(R.string.hello);
CharSequence cha = getResources().getText(R.string.app_name);
```

（2）数组。数组存储在“/res/values/arrays.xml”文件中，获取数组内容的代码如下：

```
String strs[] = getResources().getStringArray(R.array.flavors);
```

（3）颜色。颜色存储在“/res/values/colors.xml”文件中，格式为“<color name="text_color">#F00</color>”，颜色是一个整数，需要通过 R.color 来获取。

（4）尺寸值。尺寸值存储在“/res/values/dimens.xml”文件中，获取尺寸值的代码如下：

```
float myDimen = getResources().getDimension(R.dimen.dimen 标签 name 属性的名字);
```

（5）图像和图标。图像和图标存储在“/res/drawable/drawables.xml”中，获取图像和图标的代码中如下：

```
ColorDrawable myDraw = (ColorDrawable)getResources().getDrawable(R.drawable.red_rect);
```

14.4　开发实践：智能电表日志记录

14.4.1　开发设计

根据任务的开发需求，智能电表日志要包括能耗数据（用电量）。本任务使用 SharedPreferences 来保存当天的能耗数据，当再次打开应用程序时，可以显示之前保存的能耗数据。智能电表日志记录界面如图 14.2 所示，界面布局结构如图 14.3 所示。

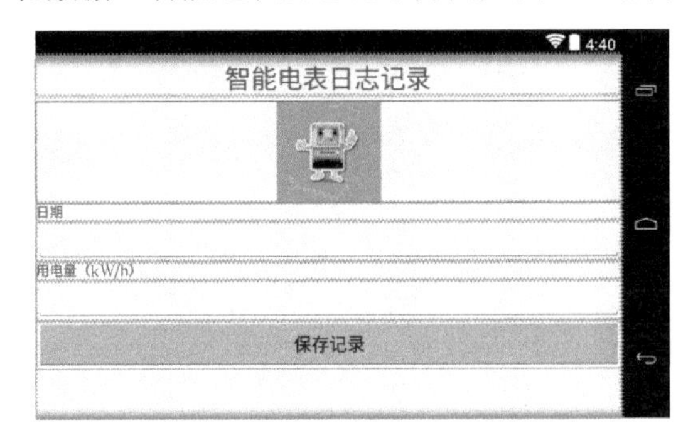

图 14.2　智能电表日志记录界面

LinearLayout (vertical)
Ab **text_header** (TextView) - "@string/app_name"
imageVie1 (ImageView)
Ab TextView - "日期"
Ab **name** (EditText)
Ab TextView - "用电量（kw/h）"
Ab **height** (EditText)
button - "保存记录"

图 14.3　界面布局结构

本任务主要编写 MainActivity.java 程序文件与 activity_main.xml 布局文件。项目目录结构如图 14.4 所示。

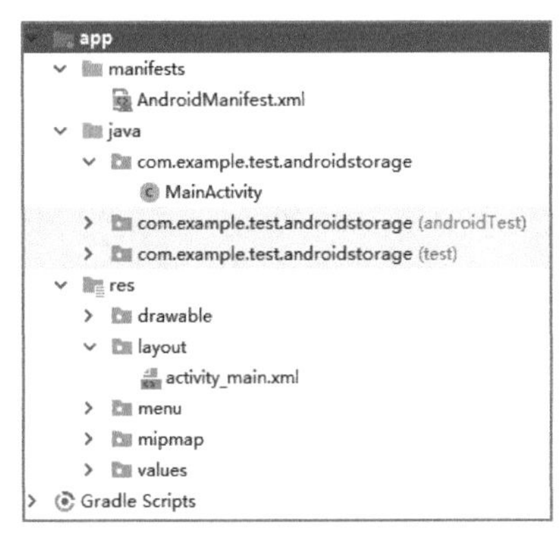

图 14.4　项目目录结构

项目开发步骤如下：

（1）编写 activity_main.xml 布局文件，设置界面显示的控件布局。

（2）编写 MainActivity.java 程序，先实现类的 save()方法，通过 getSharedPreferences()方法获取一个 SharedPreferences 对象，通过 putString()方法保存数据，通过 commit()方法进行提交数据。

（3）在 onCreate()方法中初始化 SharedPreferences 对象，在“保存记录”按钮的单击事件监听器中通过 save()方法保存用户输入的数据。

14.4.2 功能实现

智能电表日志记录界面的日期和能耗数据（用电量）通过 EditText 控件来输入，并在“用电量”下面放置一个用于保存数据的 Button 控件（即“保存记录”按钮）。输入日期和用电量后，单击“保存记录”按钮即保存数据。当再次启动应用程序后，能够显示之前输入的数据。

1. 布局文件设计

```
<?xml version="1.0" encoding="utf-8"?>
<LinearLayout xmlns:android="http://schemas.android.com/apk/res/android"
    android:layout_width="fill_parent"
    android:layout_height="fill_parent"
    android:orientation="vertical" >
    <TextView
        android:id="@+id/text_header"
        android:layout_width="fill_parent"
        android:layout_height="wrap_content"
        android:layout_marginBottom="5dp"
        android:layout_marginTop="5dp"
        android:gravity="center"
        android:text="@string/app_name"
        android:textSize="26sp" />
    <ImageView
        android:id="@+id/imageVie1"
        android:layout_width="match_parent"
        android:layout_height="100dp"
        android:scaleType="fitCenter"
        android:src="@drawable/b11" />
    <TextView
        android:layout_width="fill_parent"
        android:layout_height="wrap_content"
        android:text="日期" />
    <EditText
        android:id="@+id/name"
        android:layout_width="fill_parent"
        android:layout_height="wrap_content"
        android:text="" />
    <TextView
        android:layout_width="fill_parent"
        android:layout_height="wrap_content"
        android:text="用电量（kW/h）" />
     <EditText
        android:id="@+id/height"
        android:layout_width="fill_parent"
        android:layout_height="wrap_content"
```

```
        android:text="" />

    <Button
        android:id="@+id/button"
        android:layout_width="match_parent"
        android:layout_height="wrap_content"
        android:text="保存记录" />
</LinearLayout>
```

2．SharedPreferences 设计

```
public class MainActivity extends Activity {
    public static final String SETTING_INFOS = "SETTING_Infos";
    public static final String DATE = "DATE";
    public static final String POWER = "POWER";
    private EditText field_date;
    private EditText field_power;
    private Button button;
    @Override
    public void onCreate(Bundle savedInstanceState) {
        super.onCreate(savedInstanceState);
        //去掉 TitleBar
        requestWindowFeature(Window.FEATURE_NO_TITLE);
        setContentView(R.layout.activity_main);
        //find View
        field_date = (EditText) findViewById(R.id.name);
        field_power = (EditText) findViewById(R.id.height);

        //获取一个 SharedPreferences 对象
        SharedPreferences setting = getSharedPreferences(SETTING_INFOS, 0);
        String name = setting.getString(DATE, "");
        String height = String.valueOf(setting.getFloat(POWER, (float) 0.0));
        //Set value
        field_date.setText(name);
        field_power.setText(height);
        button = findViewById(R.id.button);
        button.setOnClickListener(new View.OnClickListener() {
            @Override
            public void onClick(View view) {
                save();
                Toast.makeText(MainActivity.this, "日志保存成功，请关闭后 APP，
                                        重新打开查看上次日志。", Toast.LENGTH_SHORT).show();
            }
        });
    }
    //将日期和用电量保存进去
    private void save() {
```

```
            //首先获取一个 SharedPreferences 对象
            SharedPreferences setting = getSharedPreferences(SETTING_INFOS, 0);
            setting.edit().putString(DATE, field_date.getText().toString())
                        .putFloat(POWER, Float.parseFloat(field_power.getText().toString())).commit();
        }
        @Override
        protected void onStop() {
            super.onStop();
        }
    }
```

14.5 任务验证

在 Android Studio 开发环境中打开本任务的例程，编译通过后运行程序，如图 14.5 所示。

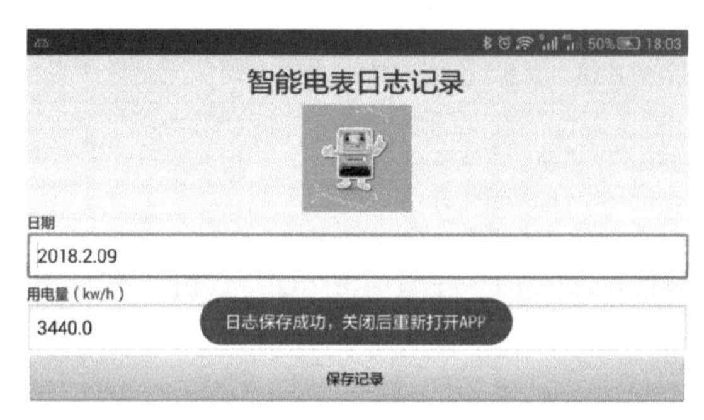

图 14.5　程序运行界面

14.6 开发小结

本任务介绍了 Android 系统中 SharedPreferences 和文件存储，读者可以利用本任务介绍的内容来完成智能电表日志的记录。

14.7 思考与拓展

（1）SharedPreferences 是用什么方式将数据保存在手机中的？

（2）请尝试为智能电表日志增加一项数据保存的功能。

任务 15

光照度记录的查询

本任务主要介绍 SQLite 数据库的操作。通过本任务的学习，读者可以掌握 SQLite 数据库的创建及其基本的数据操作方法，实现基于 SQLite 数据库的光照度记录查询。

15.1 开发场景：如何显示 SQLite 数据库中的光照度记录

假设有一个 SQLite 数据库，里面保存了光照度记录。如何在 Android 界面查询光照度记录呢？本任务主要实现光照度记录的查询。

15.2 开发目标

（1）知识目标：熟悉 SQLite 数据库的创建及其基本的数据操作方法。
（2）技能目标：掌握 SQLite 数据库的基本数据操作方法。
（3）任务目标：实现基于 SQLite 数据库的光照度记录的查询。

15.3 原理学习：SQLite 数据库的创建及其基本的数据操作方法

15.3.1 SQLite 数据库

1. SQLite 数据库简介

SQLite 是一款可以内置到移动设备上的轻量关系型数据库，多用于嵌入式系统中。SQLite 数据库由 SQL 编译器、内核、后端及附件四个部分组成。通过虚拟机和虚拟数据库引擎，可以使调试、修改和扩展 SQLite 数据库的内核变得更加方便。

SQLite 数据库和其他数据库最大的不同就是对数据类型的支持，SQLite 数据库支持 null、integer、real、text 和 blob 等数据类型。虽然 SQLite 数据库支持的类型只有 5 种，但实际上 SQLite 数据库也接收 varchar、char 和 decimal 等数据类型，只不过在运算或保存时会转变成上述的 5 种数据类型。

2．在命令行窗口中以手动的方式创建 SQLite 数据库

SQLite3 是 SQLite 数据库自带的一个基于命令行的 SQL 命令执行工具，并可以显示命令执行结果。SQLite3 已经集成到了 Android 系统中，用户在 Windows 的命令行窗口中输入“sqlite3”即可启动 SQLite3 来进行 SQLite 数据库的相关操作。在命令行窗口中以手动的方式创建 SQLite 数据库的步骤如下：

（1）下载 sqlite-tools-win32-*.zip 和 sqlite-dll-win32-*.zip（64 位的操作系统请选择 64 位版本）压缩文件。解压两个压缩包，分别把其中的 sqlite3.def、sqlite3.dll、sqlite.exe 文件复制到“C:\sqlite”目录下，在命令行窗口运行“sqlite3”，如图 15.1 所示。

```
C:\WINDOWS\system32\cmd.exe - sqlite3
2019/11/25  14:34         1,806,671 sqlite-tools-win32-x86-3300100.zip
2019/10/11  06:57             6,032 sqlite3.def
2019/10/11  06:57         1,899,008 sqlite3.dll
2019/10/11  19:28           953,856 sqlite3.exe
2019/10/11  19:28         2,019,328 sqlite3_analyzer.exe
               7 个文件      7,996,641 字节
               4 个目录 44,156,612,608 可用字节

C:\sqlite>sqlite3
SQLite version 3.30.1 2019-10-10 20:19:45
Enter ".help" for usage hints.
Connected to a transient in-memory database.
Use ".open FILENAME" to reopen on a persistent database.
sqlite>
```

图 15.1　在命令行窗口运行“sqlite3”

（2）在命令行窗口中输入“sqlite3 DBstudent.db”即可创建数据库，如图 15.2 所示。

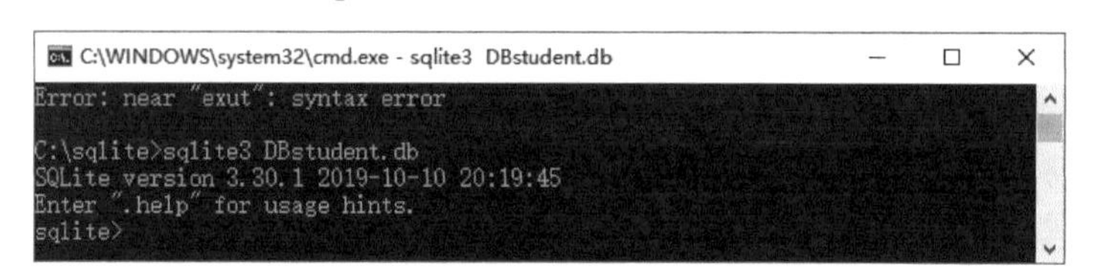

图 15.2　在命令行窗口中创建数据库

（3）在命令行窗口中输入“create table Student(id integer primary key,name text,age integer);”即可在数据库中创建一张表，接着在命令行窗口中输入“insert into Student(id,name,age) values(1,'Shen',21);”来存入两条数据，最后通过“select * from Student;”进行查询显示，如图 15.3 所示。

```
C:\WINDOWS\system32\cmd.exe - sqlite3 DBstudent.db
Enter ".help" for usage hints.
sqlite> create table Student(id integer primary key,name text,age integer);
sqlite> insert into Student(id,name,age) values(1,'Shen',21);
sqlite> insert into Student(id,name,age) values(2,'Wu',22);
sqlite> select * from Student;
1|Shen|21
2|Wu|22
sqlite>
```

图 15.3　创建表、存入数据并进行查询

（4）通过命令“update Student set name='shenhan' where id=1;”可以修改表中的记录，如

图 15.4 所示。

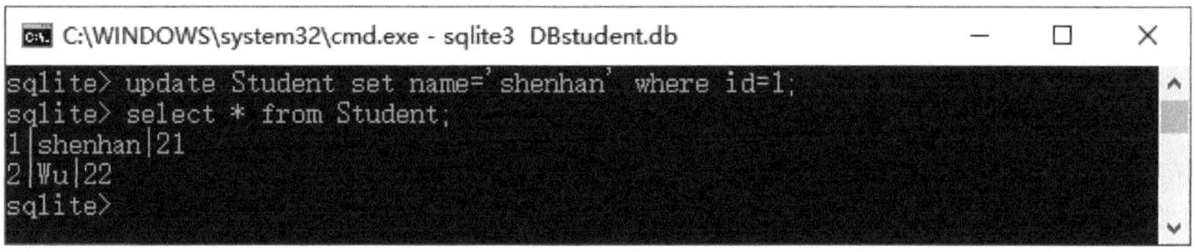

```
C:\WINDOWS\system32\cmd.exe - sqlite3  DBstudent.db
sqlite> update Student set name='shenhan' where id=1;
sqlite> select * from Student;
1|shenhan|21
2|Wu|22
sqlite>
```

图 15.4　修改表中的记录

（5）通过命令“delete from Studnet where id=1;”可以删除表中的记录，如图 15.5 所示。

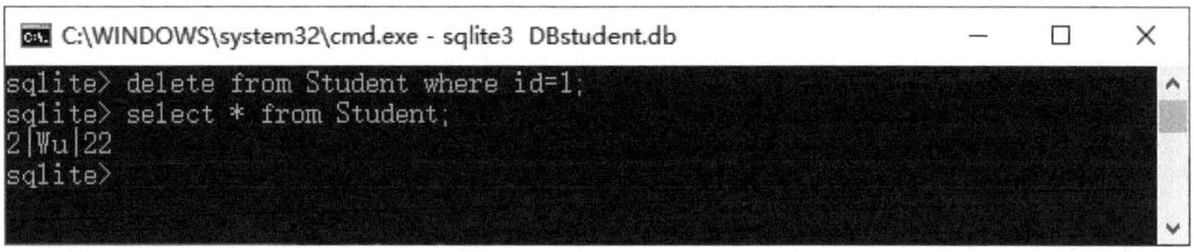

```
C:\WINDOWS\system32\cmd.exe - sqlite3  DBstudent.db
sqlite> delete from Student where id=1;
sqlite> select * from Student;
2|Wu|22
sqlite>
```

图 15.5　删除表中的记录

（6）通过命令“.tables”可以显示表和视图，通过命令“.databases”可以列出当前数据文件中的数据库，如图 15.6 所示。

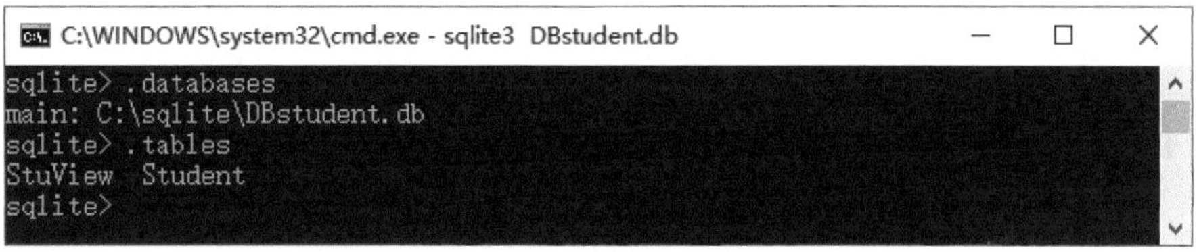

```
C:\WINDOWS\system32\cmd.exe - sqlite3  DBstudent.db
sqlite> .databases
main: C:\sqlite\DBstudent.db
sqlite> .tables
StuView  Student
sqlite>
```

图 15.6　显示表和视图，以及列出当前数据文件中的数据库

3. 代码建库

动态建立 SQLite 数据库是比较常用的方法。在应用程序运行中，当需要进行数据库操作时，应用程序会尝试打开数据库。如果数据库不存在，应用程序就会先自动建立数据库，再打开数据库。

在 Android 应用程序中创建 SQLite 数据库的方法有两种：一是使用继承自 SQLiteOpenHelper 的自定义类；二是通过 openOrCreateDatabases()方法创建数据库。

（1）使用继承自 SQLiteOpenHelper 的自定义类来创建数据库。在 Android 应用程序中使用 SQLite 数据库时，必须先创建数据库，然后创建表、索引并添加数据。Android 提供的 SQLiteOpenHelper 类可用于创建一个数据库，只要继承 SQLiteOpenHelper 类就可以轻松地创建数据库。

在继承 SQLiteOpenHelper 类时，需要使用以下三个方法：

① 调用 SQLiteOpenHelper 类的构造函数：有 4 个参数，Context context 表示上下文环境（如 Activity），String name 表示数据库名字，CursorFactory cursorFactory 表示一个可选的游标工厂（通常是 Null），int version 表示正在使用的数据库版本。

② onCreate()方法：使用 SQLiteDatabase 对象作为参数，再根据需要对这个对象填充并

初始化数据，参数 SQLiteDatabase db 表示 SQLiteDatabase 对象。

③ onUpgrade()方法：它需要 3 个参数，SQLiteDatabase db 表示 SQLiteDatabase 对象，int oldVersion 表示旧版本号，int newVersion 表示新版本号。

在 Android 应用程序中创建 SQLite 数据库的步骤如下：

① 创建继承自 SQLiteOpenHelper 的自定义类 DatabaseHelper，并实现上述三个方法，代码如下：

```
public class DatabaseHelper extends SQLiteOpenHelper{
    DatabaseHelper(Context context, String name, CursorFactory cursorFactory, int version){
        super(context, name, cursorFactory, version);
    }
    @Override
    public void onCreate(SQLiteDatabase db){
        //TODO 创建 SQLite 数据库后，对 SQLite 数据库进行操作
    }
    @Override
    public void onUpgrade(SQLiteDatabase db, int oldVersion, int newVersion){
        //TODO 更改 SQLite 数据库版本的操作
    }
    @Override
    public void onOpen(SQLiteDatabase db){
        super.onOpen(db);
        //TODO 每次成功打开 SQLite 数据库后首先被执行
    }
}
```

② 获取 SQLiteDatabase 对象。根据需要改变 SQLite 数据库中的内容，调用 getReadableDatabase()方法或 getWriteableDatabase()方法来获取 SQLiteDatabase 对象，例如：

```
db = (new DatabaseHelper(getContext())).getWritableDatabase();
```

上面的代码会返回一个 SQLiteDatabase 对象，使用这个对象就可以查询或者修改 SQLite 数据库。当完成了对 SQLite 数据库的操作（如 Activity 已经关闭）时，需要调用 SQLiteDatabase 的 close()方法来关闭 SQLite 数据库。

（2）调用 openOrCreateDatabase()方法来创建 SQLite 数据库。android.content.Context 中提供的 openOrCreateDatabase()方法可用于创建 SQLite 数据库，例如：

```
db = context.openOrCreateDatabase(String DATABASE_UNAME, int Context, MODE_PRIVATE，null);
```

15.3.2 SQLite 数据库的操作

Android 将 SQLite 数据库的操作封装在一个类中，只要调用这个类就可以完成 SQLite 数据库的添加、更新、删除和查询等操作。下面介绍在 SQLite 数据库中创建表和索引、给表添加数据、查询 SQLite 数据库等操作。

1. 创建表和索引

为了创建表和索引，需要调用 SQLiteDatabase 的 execSQL()方法。如果没有异常，该方法是没有返回值的，例如：

```
db.execSQL("CREATE TABLE mytable(_id INTEGER PRIMARY KEY  AUTOINCREMENT, title TEXT,
value REAL);");
```

上述语句创建了一张名为 mytable 的表，表中有一个名为_id 的列，并且是主键列，列值是自动增长的整数，另外还有两列，即 title 和 value。SQLite 数据库会自动为主键列创建索引。通常，在第一次创建数据库时会同时创建表和索引。

另外，SQLiteDatabase 类提供了一个重载后的 execSQL(String sql, Object[] bindArgs)方法，该方法支持占位符参数（?），例如：

```
SQLiteDatabase db = …;
db.execSQL("insert into person(name, age) values(?,?)", new Object[]{"Tom",4});
db.close();
```

execSQL()方法的第一个参数为 SQL 语句，第二个参数为 SQL 语句中占位符参数的值，参数值在数组中的顺序要和占位符的位置相对应。

2. 给表添加数据

给数据库中的表添加数据的方法有以下两种：

（1）通过 execSQL()方法执行 INSERT、UPDATE、DELETE 等语句来更新表的数据。execSQL()方法适用于不返回结果的 SQL 语句。例如：

```
db.execSQL("INSERT INTO widgets(name, inventory) VALUES('Sprocket',5)");
```

（2）通过 SQLiteDatabase 的 insert()、update()、delete()方法来更新表的数据，这些方法把 SQL 语句作为参数来使用。

① insert()方法。insert()方法用于添加数据，各个字段的数据使用 ContentValues 来存放。ContentValues 与 Map 类似，但与 Map 相比，它提供了 put(String key Xxx value)方法和 getAsXxx(String key)方法来存取数据，其中，Xxx 表示常用的数据类型，key 为字段名称，value 为字段值。例如：

```
SQLiteDatabase db = databaseHelper.getWritableDatabase();
ContentValues values = new ContentValues();
values.put("name","Tom");
values.put("age",4);
long rowed = db.insert("person", null, values);
//返回新添加记录的行号，与主键 ID 无关
```

不管 insert()方法中的第三个参数是否包含数据，执行该方法后必然会添加一条记录。如果第三个参数为空，则会添加一条除主键之外其他字段值为 Null 的记录。

② update()方法。update()方法有 4 个参数，分别是表名、表示列名和值的 ContentValues 对象、可选的 WHERE 条件，以及可选的填充 WHERE 条件的字符串（这些字符串会替换

WHERE 条件中的“？”）。update()方法可以根据条件来更新指定列的值，通过 execSQL()方法可以达到同样的目的。WHERE 条件及其参数和其他 SQL API 类似。例如：

```
String[] parms=new String[] {"this is a string"};
db.update("Widgets", replacements, "name=?", parms);
```

③ delete()方法。delete()方法的使用和 update()类似，该方法有 3 个参数，分别是表名、可选的 WHERE 条件和可选的填充 WHERE 条件的字符串。例如：

```
db.delete("person","personid<?", new String[] {"2"});
db.close();
```

3. 查询数据库

在 Android 中，SQLite 数据库查询结果的返回值并不是数据集合的完整备份，而是返回数据集的指针，这个指针就是 Cursor 类。Cursor 类可以在查询的数据集合中以多种方式移动，并能够获得数据集合的属性名称和序号。

SQLite 数据库的查询有两种方法，分别是通过 rawQuery()方法调用 SELECT 语句以及通过 query()方法构建一个查询。

（1）通过 rawQuery()方法调用 SELECT 语句来查询 SQLite 数据库。例如：

```
Cursor c=db.rawQuery("SELECT name FROM sqlite_master WHERE type='table' AND name='mytable'", null);
```

rawQuery()方法的第一个参数为 SELECT 语句；第二个参数为 SELECT 语句中占位符参数的值，如果 SELECT 语句没有使用占位符，则该参数可以设置为 null。带占位符参数的 SELECT 语句使用示例如下：

```
Cursor cursor = db.rawQuery("select * from person where name like ? and age=?", new String[]{"%Tom%", "4"});
```

上面语句的返回值是一个 Cursor 对象，这个对象的方法可以迭代查询结果。如果查询是动态的，使用 rawQuery()方法就会变得非常复杂。如果在程序编译时不能确定需要查询的列，则使用 query()方法会方便很多。

（2）使用 query()方法构建一个查询。query()函数的语法如下：

```
Cursor android.database.sqlite.SQLiteDatabase.query(String table, String[] columns, String selection, String[] selectionArgs, String groupBy, String having, String orderBy, String limit)
```

query()方法的参数说明如表 15.1 所示。

表 15.1 query()方法的参数说明

参　数	说　明
String table	表名称
String[] columns	返回的属性列名称
String selection	查询条件语句
String[] selectionArgs	如果在查询条件中使用问号，则需要定义占位符的具体内容

续表

参　数	说　明
String groupBy	分组方式
String having	定义组的过滤器
String orderBy	排序方式
String limit	指定偏移量和获取的记录数

例如：

```
SQLiteDatabase db = databaseHelper.getWritableDatabase();
Cursor cursor = db.query("person", new String[]{"personid,name,age"}, "name like ?", new String[]
{"%Tom%"}, null, null, "personid desc", "1,2");
while (cursor.moveToNext()){
    int personid = cursor.getInt(0);           //获取第一列的值，索引是从 0 开始的
    String name = cursor.getString(1);         //获取第二列的值
    int age = cursor.getInt(2);                //获取第三列的值
}
cursor.close();
db.close();
```

在 SQLite 数据库中使用游标时，结果会返回一个 Cursor 对象。Cursor 类常用方法如表 15.2 所示。

表 15.2　Cursor 类的常用方法

方　法	说　明
moveToFirst	将指针移动到第一条数据上
moveToNext	将指针移动到下一条数据上
moveToPrevious	将指针移动到上一条数据上
getCount	获取集合的数据数量
getColumnIndexOrThrow	返回指定属性名称的序号，如果属性不存在则抛出异常
getColumnName	返回指定序号的属性名称
getColumnNames	返回属性名称的字符串数组
getColumnIndex	根据属性名称返回序号
moveToPosition	将指针移动到指定的数据上
getPosition	返回当前指针的位置
getString、getInt 等	获取给定字段当前记录的值
requery	重新执行查询得到游标
close	释放游标资源

15.3.3 SQLite 简单示例

本节通过一个简单的示例来介绍 SQLite 数据库和表的创建，以及 SQLite 数据库的增、删、改、查等操作。该示例有两个类：调试类 com.example.testSQLite 和数据库辅助类 com.example. testSQLiteDb。

（1）StuDBHelper.java 的代码如下：

```
public class StuDBHelper extends SQLiteOpenHelper {
    private static final String TAG = "TestSQLite";
    //必须有构造函数
    public StuDBHelper(Context context, String name, CursorFactory factory, int version) {
        super(context, name, factory, version);
    }
    //当第一次创建 SQLite 数据库时调用 onCreate()方法
    public void onCreate(SQLiteDatabase db) {
        String sql = "create table student (id varchar(10),   name varchar(10), age integer);";
        //输出创建 SQLite 数据库的日志信息
        Log.i(TAG, "create Database------------->");
        //execSQL()方法用于执行 SQL 语句
        db.execSQL(sql);
    }
    //当更新 SQLite 数据库时执行调用 onUpgrade()方法
    public void onUpgrade(SQLiteDatabase db, int oldVersion, int newVersion) {
        //输出更新 SQLite 数据库的日志信息
        Log.i(TAG, "update Database------------->");
    }
}
```

（2）SQLiteActivity.java 的代码如下：

```
public class SQLiteActivity extends Activity {
    private Button createBtn;
    private Button insertBtn;
    private Button queryBtn;
    private Button deleteBtn;
private Button ModifyBtn;
private StuDBHelper dbHelper;
    @Override
    public void onCreate(Bundle savedInstanceState) {
        super.onCreate(savedInstanceState);
        setContentView(R.layout.main);
        initViews();
        setListener();
    }
    //通过 findViewById()方法获得 Button 对象的方法
    private void initViews() {
```

```
        createBtn = (Button) findViewById(R.id.createDatabase);
        insertBtn = (Button) findViewById(R.id.insert);
        ModifyBtn = (Button) findViewById(R.id.update);
        queryBtn = (Button) findViewById(R.id.query);
        deleteBtn = (Button) findViewById(R.id.delete);
    }
    //为按钮注册监听的方法
    private void setListener() {
        createBtn.setOnClickListener(new CreateListener());
        insertBtn.setOnClickListener(new InsertListener());
        ModifyBtn.setOnClickListener(new ModifyListener());
        queryBtn.setOnClickListener(new QueryListener());
        deleteBtn.setOnClickListener(new DeleteListener());
    }
    //创建 SQLite 数据库的方法
    class CreateListener implements OnClickListener {
        @Override
        public void onClick(View v) {
            //创建 StuDBHelper 对象
            dbHelper = new StuDBHelper(SQLiteActivity.this, "stu_db", null, 1);
            //得到一个可读的 SQLiteDatabase 对象
            SQLiteDatabase db = dbHelper.getReadableDatabase();
        }
    }
    //插入数据的方法
    class InsertListener implements OnClickListener {
        @Override
        public void onClick(View v) {
            SQLiteDatabase db = dbHelper.getWritableDatabase();
            //生成 ContentValues 对象，key 为列名，value 为想插入的值
            ContentValues cv = new ContentValues();
            cv.put("id", 1);
            cv.put("sname", "xiaoming");
            cv.put("sage", 21);
            cv.put("ssex", "male");
            //调用 insert()方法将数据插入 SQLite 数据库
            db.insert("stu_table", null, cv);
            db.close();
        }
    }
    //查询数据的方法
    class QueryListener implements OnClickListener {
        @Override
        public void onClick(View v) {
            StuDBHelper dbHelper = new StuDBHelper(SQLiteActivity.this,"stu_db", null, 1);
            SQLiteDatabase db = dbHelper.getReadableDatabase();
            Cursor cursor = db.query("stu_table", new String[] { "id", "sname", "sage", "ssex" }, "id=1",
```

```
                    null, null, null, null);
            while (cursor.moveToNext()) {
                String name = cursor.getString(cursor.getColumnIndex("sname"));
                String age = cursor.getString(cursor.getColumnIndex("sage"));
                String sex = cursor.getString(cursor.getColumnIndex("ssex"));
                System.out.println("query------->" + "姓名：" + name + " " +"年龄：" + age + " " + "性别：
                                                        " + sex);
            }
            db.close();
        }
    }
    //修改数据的方法
    class ModifyListener implements OnClickListener {
        @Override
        public void onClick(View v) {
            SQLiteDatabase db = dbHelper.getWritableDatabase();
            ContentValues cv = new ContentValues();
            cv.put("sage", "23");
            String whereClause = "id=1";
            db.update("stu_table", cv, whereClause, null);
        }
    }
    //删除数据的方法
    class DeleteListener implements OnClickListener {
        @Override
        public void onClick(View v) {
            SQLiteDatabase db = dbHelper.getWritableDatabase();
            String whereClauses = "id=2";
            //调用 delete()方法删除数据
            db.delete("stu_table", whereClauses, null);
        }
    }
}
```

（3）Main.xml 的代码如下：

```
<?xml version="1.0" encoding="utf-8"?>
<LinearLayout xmlns:android="http://schemas.android.com/apk/res/android"
    android:layout_width="match_parent"
    android:layout_height="match_parent"
    android:orientation="vertical"
    android:padding="20dp">
    <Button
        android:id="@+id/createDatabase"
        android:layout_width="match_parent"
        android:layout_height="wrap_content"
        android:text="创建 SQLite 数据库" />
    <Button
```

```
        android:id="@+id/insert"
        android:layout_width="match_parent"
        android:layout_height="wrap_content"
        android:text="插入数据" />
    <Button
        android:id="@+id/update"
        android:layout_width="match_parent"
        android:layout_height="wrap_content"
        android:text="更新数据" />
    <Button
        android:id="@+id/query"
        android:layout_width="match_parent"
        android:layout_height="wrap_content"
        android:text="查询数据" />
    <Button
        android:id="@+id/delete"
        android:layout_width="match_parent"
        android:layout_height="wrap_content"
        android:text="删除数据" />
</LinearLayout>
```

运行效果如图 15.7 所示。

图 15.7 运行效果

15.4 开发实践：光照度记录的查询

15.4.1 开发设计

本任务使用 ListActivity 作为主界面的显示框架，用来显示光照度记录；设计 DBHelper 类，专门用于操作 SQLite 数据库，如增、删、改、查等。光照度记录查询界面设计如图 15.8 所示。

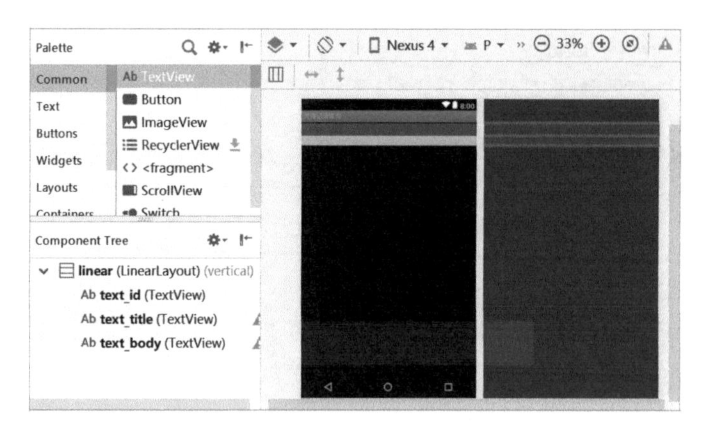

图 15.8　光照度记录查询界面设计

本任务主要编写 MainActivity.java 程序文件与 activity_main.xml 布局文件，项目文件如表 15.3 所示。

表 15.3　项目文件

文　件　名	功 能 说 明
DBHelper.java	SQLite 数据库操作类，实现 SQLite 数据库的增、删、改、查等操作
DbLUX.java	SQLite 数据库表类，定义表中数据字段
MainActivity.java	主界面程序文件，新建 SQLite 数据库对象，列表显示
SeeDetailNews.java	子界面程序文件，显示列表详细数据
activity_main.xml	主界面布局文件
content_main.xml	子界面布局文件

项目目录结构如图 15.9 所示。

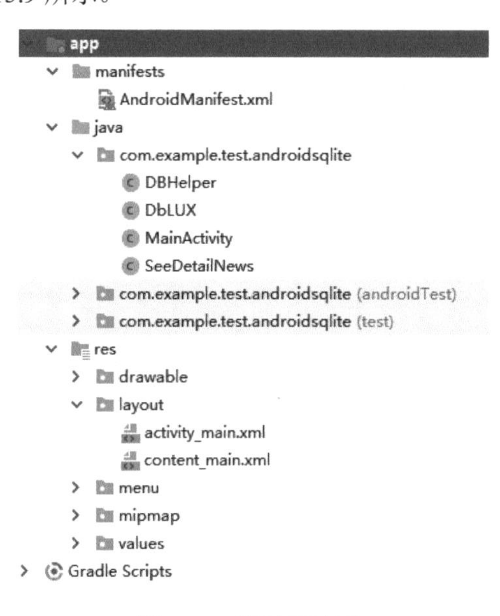

图 15.9　项目目录结构

项目开发步骤如下：

（1）编写 activity_main.xml 布局文件，设置主界面显示的控件布局。

（2）编写 content_main.xml 布局文件，设置子界面显示的控件布局。

（3）编写 DbLUX.java 程序文件，SQLite 数据库表类，用于定义表中数据字段。

（4）编写 DBHelper.java 程序文件，SQLite 数据库操作类，用于实现 SQLite 数据库的增、删、改、查等操作。

（5）编写 SeeDetailNews.java 程序文件，通过 getStringArrayExtra()方法从主界面获取的数据以列表的形式显示。

（6）编写 MainActivity.java 程序文件，实例化 DBHelper 对象，实例化一个用于数据绑定的 Adapter 类，即 SimpleCursorAdapter，通过 setListAdapter(scadaAdapter)方法绑定到 ListActivity 对象。

（7）覆写 onListItemClick()方法，在该方法中通过 helper.queryByID(objId)方法查询 SQLite 数据库，并通过 Intent 的 putExtra()方法向子界面传递数据。

15.4.2　功能实现

首先创建 Android 项目，定义界面布局文件；然后创建 SQLite 数据库操作类；最后编写主界面的程序代码，读取 SQLite 数据库中的记录并显示在主界面上。

1．布局文件设计

```
<?xml version="1.0" encoding="utf-8"?>
<LinearLayout xmlns:android="http://schemas.android.com/apk/res/android"
    android:id="@+id/linear"
    android:layout_width="fill_parent"
    android:layout_height="wrap_content"
    android:gravity="top"
    android:orientation="vertical"
    android:paddingBottom="0dp">
    <TextView
        android:id="@+id/text_id"
        android:layout_width="fill_parent"
        android:layout_height="2dp"
        android:visibility="invisible" />
    <TextView
        android:id="@+id/text_title"
        android:layout_width="fill_parent"
        android:layout_height="wrap_content"
        android:background="#009688"
        android:ellipsize="end"
        android:paddingLeft="10dp"
        android:singleLine="true"
        android:textColor="#000000"
        android:textSize="20dp"
```

```
            android:textStyle="bold" />
        <TextView
            android:id="@+id/text_body"
            android:layout_width="fill_parent"
            android:layout_height="wrap_content"
            android:background="#B2DFDB"
            android:maxLines="2"
            android:paddingLeft="10dp"
            android:textColor="#000000"
            android:textSize="15dp" />
</LinearLayout>
```

2. 布局文件设计

```
<?xml version="1.0" encoding="utf-8"?>
<LinearLayout xmlns:android="http://schemas.android.com/apk/res/android"
        android:id="@+id/linear"
        android:orientation="vertical"
        android:layout_width="fill_parent"
        android:layout_height="fill_parent">
        <TextView
            android:id="@+id/detail_title"
            android:layout_width="fill_parent"
            android:layout_height="wrap_content"
            android:paddingLeft="10dp"
            android:textSize="20dp"
            android:textColor="#000000"
            android:textStyle="bold"
            android:background="#009688" />
        <TextView
            android:id="@+id/detail_body"
            android:padding="3dp"
            android:textSize="15dp"
            android:textColor="#000000"
            android:background="#B2DFDB"
            android:layout_width="fill_parent"
            android:layout_height="wrap_content" />
        <Button
            android:id="@+id/button_close"
            android:layout_width="wrap_content"
            android:layout_height="wrap_content"
            android:paddingLeft="20dp"
            android:paddingRight="20dp"
            android:text="close"/>
</LinearLayout>
```

3．Manifest.xml 的设计

```
<?xml version="1.0" encoding="utf-8"?>
<manifest xmlns:android="http://schemas.android.com/apk/res/android"
    package="com.example.test.androidsqlite">
    <application
        android:allowBackup="true"
        android:icon="@mipmap/ic_launcher"
        android:label="@string/app_name"
        android:roundIcon="@mipmap/ic_launcher_round"
        android:supportsRtl="true"
        android:theme="@style/AppTheme">
        <activity
            android:name=".MainActivity"
            android:label="@string/app_name"
            android:theme="@style/AppTheme.NoActionBar">
            <intent-filter>
                <action android:name="android.intent.action.MAIN" />
                <category android:name="android.intent.category.LAUNCHER" />
            </intent-filter>
        </activity>
        <activity
            android:name="SeeDetailNews"
            android:theme="@style/AppTheme"
            android:screenOrientation="landscape"
            android:label="@string/app_name" />
    </application>
</manifest>
```

4．主界面设计

```
public class MainActivity extends ListActivity {
    private final String[] COLS = {LuxObject._ID, LuxObject.DATE, LuxObject.LUX};
    private final int[] IDS = {R.id.text_id, R.id.text_title, R.id.text_body};
    private DBHelper helper = null;
    @Override
    public void onCreate(Bundle savedInstanceState) {
        super.onCreate(savedInstanceState);
        helper = new DBHelper(this);
        Cursor cursor = helper.queryAllForCursor();
        if (cursor == null) {
            Toast.makeText(this, "Error!", Toast.LENGTH_SHORT).show();
        } else if (cursor.getCount() <= 0) {
            Toast.makeText(this, "no data exists.", Toast.LENGTH_SHORT).show();
        }
        /*实例化一个用于数据绑定的 Adapter 类，即 SimpleCursorAdapter。该类的构造函数
          SimpleCursorAdapter()中的第四个参数表示 Cursor 对象中的字段名，第五个参数表示要将该
```

```
                字段的值赋给参数指定的组件*/
            SimpleCursorAdapter scadaAdapter = new SimpleCursorAdapter(this, R.layout.main, cursor, COLS,
IDS);
            setListAdapter(scadaAdapter);
        }
        @Override
        protected void onDestroy() {
            super.onDestroy();
            helper.cleanup();
            helper = null;
        }
        @Override
        protected void onListItemClick(ListView l, View v, int position, long id) {
            super.onListItemClick(l, v, position, id);
            TextView tView = (TextView) ((LinearLayout) v).getChildAt(0);
            long objId = Long.parseLong(tView.getText().toString());      //get the id of this record
            LuxObject obj = helper.queryByID(objId);//查询记录
            Intent i = new Intent(SQLiteDemo.this, SeeDetailNews.class);//查看详细信息
            i.putExtra("extra_news", obj.toStrings());
            startActivity(i);
        }
    }
```

5. SQLite 数据库操作类设计

```
    public class DBHelper {
        private static final String[] COLS = new String[]{
            LuxObject._ID, LuxObject.DATE, LuxObject.LUX, LuxObject.RECORDER
        };
        //SQLite 数据库操作类的 API 可完成数据库的增、删、改、查等操作
        private SQLiteDatabase db;
        //打开数据库
        private DBOpenHelper dbOpenHelper;
        public DBHelper(Context context){
            this.dbOpenHelper = new DBOpenHelper(context);
            establishDb();
        }
        private void establishDb(){
            if(this.db == null){
                this.db = this.dbOpenHelper.getWritableDatabase();
            }
        }
        public void cleanup(){
            if(this.db != null){
                this.db.close();
                this.db = null;
            }
```

```
        }
        //添加记录
        public long Insert(LuxObject luxObj){
            ContentValues values = new ContentValues();
            values.put(LuxObject.DATE, luxObj.date);
            values.put(LuxObject.LUX, luxObj.lux);
            values.put(LuxObject.RECORDER, luxObj.recorder);
            return this.db.insert(DBOpenHelper.TABLE_NAME, null, values);
        }
        //更新记录
        public long update(LuxObject luxObj){
            ContentValues values = new ContentValues();
            values.put(LuxObject.DATE, luxObj.date);
            values.put(LuxObject.LUX, luxObj.lux);
            values.put(LuxObject.RECORDER, luxObj.recorder);
            return this.db.update(DBOpenHelper.TABLE_NAME, values, LuxObject._ID + "=" + luxObj.id,
null);
        }
        //删除记录
        public int delete(long id){
            return this.db.delete(DBOpenHelper.TABLE_NAME, LuxObject._ID + "=" + id, null);
        }
        public int delete(String title){
            return this.db.delete(DBOpenHelper.TABLE_NAME, LuxObject.DATE + "like " + title, null);
        }
        //查询记录
        public LuxObject queryByID(long id){
            Cursor cursor = null;
            LuxObject obj = null;
            try {
                cursor = this.db.query(DBOpenHelper.TABLE_NAME,COLS, LuxObject._ID + "=" + id,
                                                                        null, null, null,
null);
                if(cursor.getCount() > 0){
                    cursor.moveToFirst();
                    obj = new LuxObject();
                    obj.id = cursor.getLong(0);
                    obj.date = cursor.getString(1);
                    obj.lux = cursor.getString(2);
                    obj.recorder = cursor.getString(3);
                }
            } catch (SQLException e) {
                Log.v("aaaa", "aaaa->queryByID.SQLException");
            } finally{
                if(cursor != null && !cursor.isClosed()){
                    cursor.close();
                }
```

```
        }
        return obj;
    }
    public List<LuxObject> queryByTitleForList(String title){
        ArrayList<LuxObject> list = new ArrayList<LuxObject>();
        Cursor cursor = null;
        LuxObject obj = null;
        try {
            cursor = this.db.query(true, DBOpenHelper.TABLE_NAME,COLS, LuxObject.DATE +
                                        " like '%" + title + "%'", null, null, null, null, null);
            int count = cursor.getCount();
            cursor.moveToFirst();
            for(int i=0;i<count;i++){
                obj = new LuxObject();
                obj.id = cursor.getLong(0);
                obj.date = cursor.getString(1);
                obj.lux = cursor.getString(2);
                obj.recorder = cursor.getString(3);
                list.add(obj);
                cursor.moveToNext();
            }
        } catch (SQLException e) {
            Log.e("aaaa", "aaaa->queryByTitle.SQLException");
            e.printStackTrace();
        }finally{
            if(cursor != null && !cursor.isClosed())
            cursor.close();
        }
        return list;
    }
    public List<LuxObject> queryAllForList(){
        ArrayList<LuxObject> list = new ArrayList<LuxObject>();
        Cursor cursor = null;
        LuxObject obj = null;
        try {
            cursor = this.db.query(DBOpenHelper.TABLE_NAME,COLS, null, null, null, null, null);
            int count = cursor.getCount();
            cursor.moveToFirst();
            for(int i=0;i<count;i++){
                obj = new LuxObject();
                obj.id = cursor.getLong(0);
                obj.date = cursor.getString(1);
                obj.lux = cursor.getString(2);
                obj.recorder = cursor.getString(3);
                list.add(obj);
                cursor.moveToNext();
            }
```

```
        } catch (SQLException e) {
            Log.e("aaaa", "aaaa->queryByTitle.SQLException");
            e.printStackTrace();
        }finally{
            if(cursor != null && !cursor.isClosed())
            cursor.close();
        }
        return list;
    }
    public Cursor queryByTitleForCursor(String title){
        Cursor cursor = null;
        try {
            cursor = this.db.query(DBOpenHelper.TABLE_NAME, COLS, null, null, null, null, null);
        } catch (SQLException e) {
            Log.e("aaaa", "aaaa->queryByTitle.SQLException");
            e.printStackTrace();
        }finally{
        }
        return cursor;
    }
    public Cursor queryAllForCursor(){
        Cursor cursor = null;
        try {
            cursor = this.db.query(DBOpenHelper.TABLE_NAME, COLS, null, null, null, null, null);
        } catch (SQLException e) {
            Log.e("aaaa", "aaaa->queryByTitle.SQLException");
            e.printStackTrace();
        }
        return cursor;
    }
    //SQLiteOpenHelper 类用来打开或创建 SQLite 数据库
    private static class DBOpenHelper extends SQLiteOpenHelper{
        private static final String DB_NAME = "db_lux";
        private static final String TABLE_NAME = "lux_table";
        private static final int DB_VERSION = 1;
        private static final String CREATE_TABLE = "create table "+TABLE_NAME+" ("+
                        LuxObject._ID+" integer primary key," +LuxObject.DATE + " text, "+
                         LuxObject.LUX +" text, "+ LuxObject.RECORDER +" text)";
        private static final String DROP_TABLE = "drop table if exists "+TABLE_NAME;
        public DBOpenHelper(Context context){
            super(context, DB_NAME, null, DB_VERSION);
        }
        public DBOpenHelper(Context context, String name, CursorFactory factory,int version) {
            super(context, name, factory, version);
        }
        @Override
        //创建 SQLite 数据库并进行初始化
```

```
public void onCreate(SQLiteDatabase db) {
    //TODO Auto-generated method stub
    try {
        db.execSQL(CREATE_TABLE);
    } catch (SQLException e) {
        //TODO: handle exception
        Log.v("aaaa", "aaaa->Database created failed.");
    }
    //SQLite 数据库初始化
    saveSomeDatas(db, getData());
}
@Override
public void onUpgrade(SQLiteDatabase db, int oldVersion, int newVersion) {
    //TODO Auto-generated method stub
    Log.v("aaaa", "aaaa->onUpgrade, oldVersion = " + oldVersion + ", newVersion = " +
                                                    newVersion);
    try {
        db.execSQL(DROP_TABLE);
    } catch (SQLException e) {
        //TODO: handle exception
        Log.v("aaaa", "aaaa->Database created failed.");
    }
    onCreate(db);
}
private void saveSomeDatas(SQLiteDatabase db, List<Map<String, String>> value){
    ContentValues values = null;
    Map<String, String> map = null;
    while(value.size() > 0){
        map = value.remove(0);
        values = new ContentValues();
        values.put(LuxObject.DATE, map.get(LuxObject.DATE));
        values.put(LuxObject.LUX, map.get(LuxObject.LUX));
        values.put(LuxObject.RECORDER, map.get(LuxObject.RECORDER));
        db.insert(TABLE_NAME, null, values);
    }
}
private List<Map<String, String>> getData(){
    List<Map<String, String>> list = new ArrayList<Map<String,String>>();
    Map<String, String> map1 = new HashMap<String, String>();
    map1.put(LuxObject.DATE, "2018 年 02 月 20 日");
    map1.put(LuxObject.LUX, "156.5");
    map1.put(LuxObject.RECORDER, "Zhang San");
    list.add(map1);
    Map<String, String> map2 = new HashMap<String, String>();
    map2.put(LuxObject.DATE, "2018 年 02 月 21 日");
    map2.put(LuxObject.LUX, "126.8");
    map2.put(LuxObject.RECORDER, "Li Si");
```

```
                list.add(map2);
                Map<String, String> map3 = new HashMap<String, String>();
                map3.put(LuxObject.DATE, "2018 年 02 月 22 日");
                map3.put(LuxObject.LUX, "89.9");
                map3.put(LuxObject.RECORDER, "Zhang San");
                list.add(map3);
                Map<String, String> map4 = new HashMap<String, String>();
                map4.put(LuxObject.DATE, "2018 年 02 月 23 日");
                map4.put(LuxObject.LUX, "129.5");
                map4.put(LuxObject.RECORDER, "Li Si");
                list.add(map4);
                Map<String, String> map5 = new HashMap<String, String>();
                map5.put(LuxObject.DATE, "2018 年 02 月 24 日");
                map5.put(LuxObject.LUX, "168.3");
                map5.put(LuxObject.RECORDER, "Li Si");
                list.add(map5);
                return list;
            }
        }
    }
```

15.5　任务验证

在 Android Studio 开发环境中打开任务的例程，编译通过后运行程序，运行效果如图 15.9 所示。

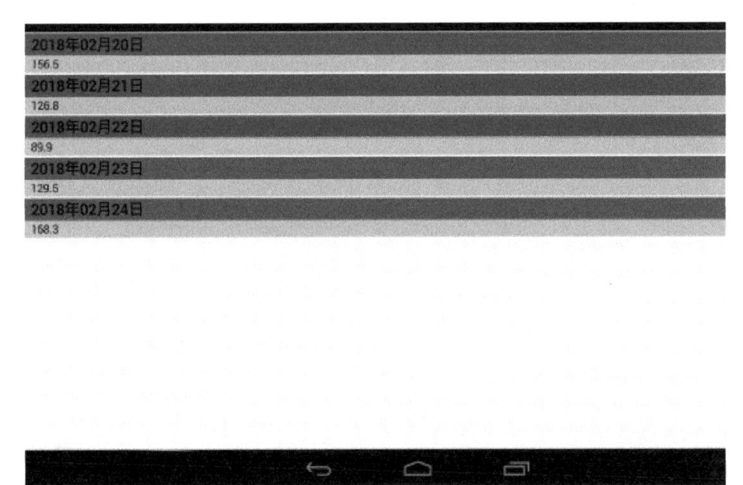

图 15.9　程序运行效果

单击界面中的一个记录会显示该记录的详细信息，如图 15.10 所示。

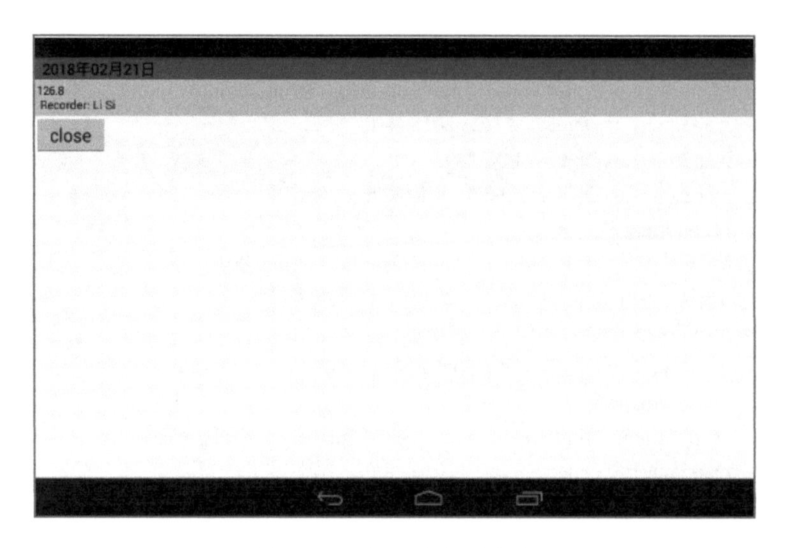

图 15.10　显示详细记录信息

15.6　开发小结

本任务主要介绍 SQLite 数据库基本的数据操作方法。通过本任务的学习，读者可以掌握 SQLite 数据库的创建以及基本的数据操作方法，实现基于 SQLite 数据库的光照度记录的查询。

15.7　思考与拓展

（1）请谈谈 SQLite 数据库和 SharedPreferences 的区别。

（2）请尝试在 SQLite 数据库中增加一条光照度记录。

任务16

智能医疗仪表图形的动态显示

本任务主要介绍动态图形的绘制和图形特效的实现。通过本任务的学习，读者可以掌握动态图形绘制的基本思路、图形特效的实现方法，从而实现智能医疗仪表图形的动态显示。

16.1 开发场景：智能医疗仪表图形动态显示的重要性及实用性

智能医疗仪表在医疗中发挥着重要的作用。例如，智能医疗仪表中的动态血压监测显示功能，有助于早期高血压病的诊断，可以协助鉴别原发性、继发性的高血压病，指导合理用药，能够更好地预防心脑血管并发症的发生，并在一定程度上预测高血压并发症的发展。

16.2 开发目标

（1）知识目标：了解动态图形绘制类和图形特效的实现方法。
（2）技能目标：掌握动态图形绘制的基本思路以及图形特效的实现方法。
（3）任务目标：实现智能医疗仪表图形的动态显示。

16.3 原理学习：动态图形的绘制及图形特效的实现

16.3.1 动态图形的绘制

1．动态图形绘制类简介

如何实现动态图形的绘制呢？Android 中的工具类都很形象，动态图形绘制类也是如此。在实现动态图形的绘制时，首先需要画布（Canvas 类）、画笔（Paint 类）和不同的颜色（Color 类），然后进行画线或者连接路径（Path 类），借助工具（ShapeDrawable 类）可以直接画出各种图形，如圆、椭圆、矩形等，ShapeDrawable 类有 OvalShape、RectShape 等子类。

（1）Canvas 类。Canvas 类位于 android.graphics 中，该类提供了绘制各种基本图形的方法。Canvas 类的详细方法如表 16.1 所示。

表 16.1　Canvas 类的详细方法

方法名称	方法描述
drawText(String text,float x,float y,Paint paint)	以(x,y)为起始坐标，使用 Paint 类绘制文本
drawPoint(float x,float y,Paint paint)	在坐标(x,y)上使用 Paint 类画点
drawLine(float startX,float startY,float stopX,float stopY,Paint paint)	以(startX, startY)为起始坐标，以(stopX, stopY)为终止坐标，使用 Paint 类画线
drawCircle(float cx,float cy,float radius,Paint paint)	以(cx, cy)为原点，以 radius 为半径，使用 Paint 类画圆
drawOval(RectF oval,Paint paint)	使用 Paint 类画矩形 oval 的内切椭圆
DrawRect(RectF rect,Paint paint)	使用 Paint 类画矩形 rect
drawRoundRect(RectF rect,float rx,float ry,Paint paint)	画圆角矩形
clipRect(float left,float top,float right,float botton)	剪辑矩形
clipRegion(Region region)	剪辑区域

（2）Paint 类。Paint 类位于 android.graphics 中，该类用用描述图形的颜色和风格，如线宽、颜色、字体等。Paint 类的详细方法如表 16.2 所示。

表 16.2　Paint 类的详细方法

方法名称	方法描述
Paint()	构造函数，使用默认设置
setColor(int color)	设置颜色
setStrokeWidth(float width)	设置线宽
setTextAlign(Paint.Align align)	设置文字对齐
setTextSize(float textSize)	设置文字尺寸
setShader(Shader shader)	设置渐变
setAlpha(int a)	设置 Alpha 值
reset()	恢复成 Paint 类的默认设置

（3）Color 类。Color 类位于 android.graphics 中，该类定义了一些颜色变量以及颜色创建的方法，颜色一般是使用 RGB 三原色定义的。Color 类的属性如表 16.3 所示。

表 16.3　Color 类的属性

属　性	说　明	属　性	说　明
BLACK	黑色	BLUE	蓝色
CYAN	青色	DKGRAY	深灰色
GRAY	灰色	GREEN	绿色
LIGRAY	浅灰色	MAGENTA	紫色
RED	红色	TRANSPARENT	透明
WHITE	白色	YELLOW	黄色

（4）Path 类。当画一个圆时，只需要指定圆心（点）和半径就可以了。如果要画一个梯形，那么该怎么办呢？这就需要点和连线。Path 类位于 android.graphics 中，该类用来画从一个点到另一个点的连线。Path 类的详细方法如表 16.4 所示。

表 16.4　Path 类的详细方法

方 法 名 称	方 法 描 述
lineTo(float x,float y)	从上次操作结束点到指定点画线，如果没有进行过操作则起点默认为原点
moveTo(float x,float y)	移动到指定点，不画线
reset()	复位

2．动态图形绘制的基本思路

动态图形绘制的基本思路是：首先创建一个继承自 View 类或 SurfaceView 类的类并覆写 onDraw()方法，然后使用 Canvas 对象在界面上绘制不同的图形，最后使用 invalidate()方法更新界面。下面通过一个弹球实例来说明动态图形绘制的基本思路，该实例在界面上动态绘制一个弹球，当弹球触顶或者触底时自动改变方向继续运行。弹球实例步骤说明如下：

（1）创建一个 Android 项目，入口 Activity 的名称为 MainActivity。

（2）在 MainActivity 类中创建一个 MyView 内部类，该内部类实现的 Runnable 接口可支持多线程。在 onDraw()方法中定义画笔并设置画笔颜色，使用 Canvas 类的 drawCircle()方法画圆。定义一个 update()方法，用于更新坐标。定义一个继承自 Handler 类的消息处理器类 RefreshHandler，并覆写 handleMessage()方法，在该方法中处理消息，在线程的 run()方法中设置并发送消息，在构造函数中启动线程。

```
public class MyView extends View implements Runnable {
    RefreshHandler mRedrawHandler = new RefreshHandler();
    Paint p = new Paint();
    //图形当前坐标
    private int x = 20, y = 20;
    //构造函数
    public MyView(Context context, AttributeSet attrs) {
        super(context, attrs);
        p.setColor(Color.GREEN);
        p.setAntiAlias(true);
        p.setDither(true);
        //获得焦点
        setFocusable(true);
        //启动线程
        new Thread(this).start();
    }
    @Override
    public void run() {
        while (!Thread.currentThread().isInterrupted()) {
            //通过发送消息更新界面
            Message m = new Message();
```

```
                m.what = 1001;
                mRedrawHandler.sendMessage(m);
                try {
                    Thread.sleep(300);
                } catch (InterruptedException e) {
                    e.printStackTrace();
                }
            }
        }
        @Override
        protected void onDraw(Canvas canvas) {
            super.onDraw(canvas);
            canvas.drawCircle(x, y, 20, p);
        }
        //定义一个消息处理类 RefreshHandler
        class RefreshHandler extends Handler {
            @Override
            public void handleMessage(Message msg) {
                if (msg.what == 1001) {
                    MyView.this.update();
                    MyView.this.invalidate();
                }
                super.handleMessage(msg);
            }
        }
        //更新坐标
        private void update() {
            x += 5;
            y += 5;
        }
}
```

（3）在 MainActivity 的 onCreate()方法中实例化 MyView 类，并将其设置为 Activity 的内容视图。

```
public void onCreate(Bundle savedInstanceState) {
    super.onCreate(savedInstanceState);
    MyView v = new MyView(this,null);
    setContentView(v);
}
```

弹球实例运行效果如图 16.1 所示。

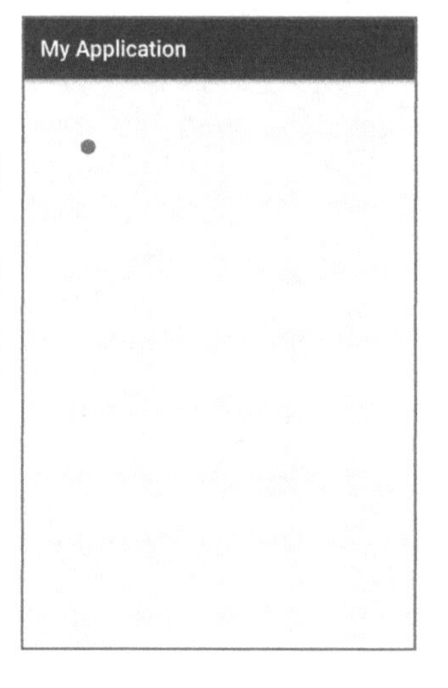

图 16.1　弹球实例运行效果

3．绘制几何图形

下面介绍如何使用前文介绍的类来绘制一些常见的几何图形，绘制步骤如下：

（1）创建一个 Activity。

（2）创建 Activity 的内部类 MyView，该类继承自 View 类。

（3）覆写 View 类的 onDraw()方法。

（4）在 onDraw()方法中创建 Paint 对象和 Path 对象，设置 Canvas 类的属性，画出各种图形。

（5）在 Activity 的 onCreate()方法中实例化 MyView，调用 Activity 的 setContentView()方法将 MyView 设置为当前视图。

```
public class MainActivity extends AppCompatActivity {
    @Override
    protected void onCreate(Bundle savedInstanceState) {
        super.onCreate(savedInstanceState);
        MyView v = new MyView(this, null);
        setContentView(v);
    }
    class MyView extends View {
        public MyView(Context context, @Nullable AttributeSet attrs) {
            super(context, attrs);
        }
        @Override
        protected void onDraw(Canvas canvas) {
            super.onDraw(canvas);
            canvas.drawColor(Color.WHITE);                    //设置 Canvas 类的颜色
            Paint paint = new Paint();                        //实例化 Paint 类
            paint.setAntiAlias(true);
            paint.setColor(Color.BLUE);                       //设置颜色
            paint.setStyle(Paint.Style.STROKE);               //设置样式
            paint.setStrokeWidth(2);                          //设置线宽
            canvas.drawCircle(80, 80, 60, paint);             //画圆
            paint.setTextSize(30);                            //设置文本大小
            canvas.drawText("圆形", 50, 180, paint);          //写文本
            canvas.drawRect(210, 30, 310, 130, paint);        //画方形
            canvas.drawText("正方形", 215, 180, paint);       //写文本
            //画三角形
            Path Tpath = new Path();                          //实例化路径
            Tpath.moveTo(400, 30);                            //移动到指定点
            Tpath.lineTo(500, 30);                            //画线
            Tpath.lineTo(450, 130);                           //画线
            Tpath.close();                                    //关闭路径
            canvas.drawPath(Tpath, paint);                    //画路径
            canvas.drawText("三角形", 400, 180, paint);       //写文本
        }
    }
}
```

绘制的几何图形如图 16.2 所示。

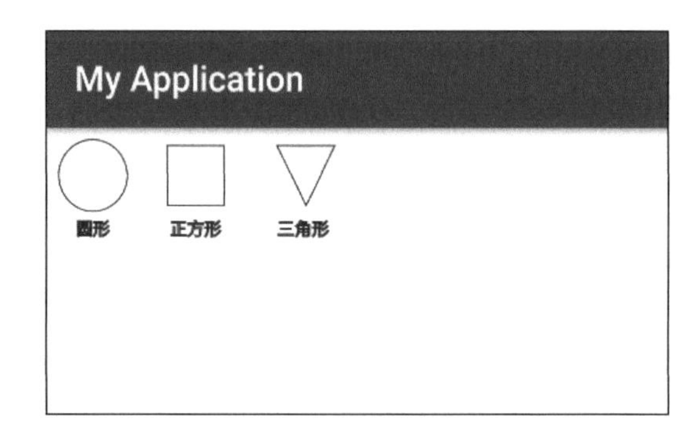

图 16.2　绘制的几何图形

16.3.2　图形特效的实现

1. 通过 Matrix 类实现图形的旋转、缩放和平移

Android 系统提供了一个 Matrix 类，该类具有一个 3×3 的矩阵坐标，通过该类可以实现图形的旋转、缩放和平移。Matrix 类的详细方法如表 16.5 所示。

表 16.5　Matrix 类的详细方法

方 法 名 称	方 法 描 述
void reset()	重置一个 Matrix 对象
void set(Matrix src)	复制一个源矩阵，与构造函数 Matrix(Matrix src) 功能一样
boolean isIdentity()	判断矩阵是否单位矩阵
void setRotate(float degrees)	以坐标（0, 0）为中心旋转指定的角度
void setRotate(float degrees, float px, float py)	以坐标（px, py）为中心旋转指定的角度
void serScale(float sx, float sy)	缩放处理
void setTranslate(float dx, float dy)	平移处理
void setSkew(float kx, float ky)	倾斜处理

下面通过具体的实例来介绍 Matrix 类的使用方法。首先自定义一个 View 类，覆写 View 类的 onDraw()方法，在该类中有一个 Bitmap 对象和 Matrix 对象；然后 Bitmap 对象从系统资源加载一张图片，在 onDraw()方法中通过 reset()方法初始化 Matrix 对象，并设置其旋转或缩放属性；接着使用 Canvas 类的 drawBitmap()方法将 Bitmap 对象重新绘制在视图中；最后通过键盘事件 onKeyDown 实现旋转属性和缩放属性的改变，调用 postInvalidate()方法重新绘制 Bitmap 对象。具体步骤如下：

（1）创建一个 Android 项目 GraphicMatrix，入口 Activity 的名称为 MainActivity。

（2）在该项目的“res/drawable/”目录下添加一张图片资源 test.jpg。

（3）在 MainActivity 中定义一个内部类 MyView，该类继承自 View 类。在 MyView 类中声明使用到的变量，在构造函数中初始化变量，覆写 onDraw()方法和 onKeyDown()方法。

代码如下：

```
class MyView extends View{
    private Bitmap bm;                                  // Bitmap 对象
    private Matrix matrix = new Matrix();               //Matrix 对象
    private float angle = 0.0f;                         //旋转角度
    private int w,h;                                    //位图的宽度和高度
    private float scale = 1.0f;                         //缩放比例
    private boolean isScale = false;                    //判断是缩放还是旋转
    //构造函数
    public MyView(Context context) {
        super(context);
        bm = BitmapFactory.decodeResource(this.getResources(), R.drawable.test);        //获得位图
        w = bm.getWidth();                              //获得位图的宽度
        h = bm.getHeight();                             //获得位图的高度
        this.setFocusable(true);                        //使当前视图获得焦点
    }
    @Override
    protected void onDraw(Canvas canvas) {
        super.onDraw(canvas);
        matrix.reset();                                 //重置 Matrix
        if(!isScale){
            matrix.setRotate(angle);                    //旋转 Matrix
        }else{
            matrix.setScale(scale, scale);              //缩放 Matrix
        }
        Bitmap bm2 = Bitmap.createBitmap(bm, 0, 0, w, h,matrix, true);
        //根据原始位图和 Matrix 创建新视图
        canvas.drawBitmap(bm2, matrix, null);           //绘制新视图
    }
    @Override
    public boolean onKeyDown(int keyCode, KeyEvent event) {
    //向左旋转
    if(keyCode == KeyEvent.KEYCODE_DPAD_LEFT){
        isScale = false;
        angle++;
        postInvalidate();
    }
    //向右旋转
    if(keyCode == KeyEvent.KEYCODE_DPAD_RIGHT){
        isScale = false;
        angle--;
        postInvalidate();
    }
    //放大
    if(keyCode == KeyEvent.KEYCODE_DPAD_UP){
        isScale =true;
        if(scale < 2.0F)
```

```
            scale += 0.1;
            postInvalidate();
        }
        //缩小
        if(keyCode == KeyEvent.KEYCODE_DPAD_DOWN){
            isScale = true;
            if(scale > 0.5F)
            scale -= 0.1;
            postInvalidate();
        }
        return super.onKeyDown(keyCode, event);
    }
```

（4）在 MainActivity 的 onCreate()方法中实例化 MyView 类，并将其设置为当前 Activity 的视图布局。

```
    public void onCreate(Bundle savedInstanceState) {
        super.onCreate(savedInstanceState);
        MyView myView = new MyView(MainActivity.this);
        setContentView(myView);                               //设置当前视图布局
    }
```

2．Bitmap 类和 BitmapFactory

位图是 Windows 操作系统的标准格式图形文件，其扩展名是.bmp 或者.dib。位图将图像看成由点（像素）组成，每个点可以由多种色彩表示，包括 2、4、8、16、24 和 32 位色彩。例如，一幅 1024×768 分辨率的 32 位色彩图片，其所占用的存储空间为 1024×768×32/8= 3072 KB。虽然 Bitmap 的图像效果较好，但位图采用的是非压缩格式，需要占用较大的存储空间，不利于在网络上传输。在 Android 中，Bitmap 类是图像处理中最重要的一种中转类。使用 Bitmap 类可以获取图像文件信息，借助 Matrix 类可进行图像的剪切、旋转、缩放等操作，最后以指定格式来保存图像文件。

通常，在构造一个类的对象时，都是通过该类的对应构造函数来实现的。但 Bitmap 类采用的是 BitmapFactory 的设计模式，通常不使用构造函数。

（1）通过 Bitmap 类的静态方法 static Bitmap createBitmap()可以构造 Bitmap 类的对象。Bitmap 类的静态方法如表 16.6 所示。

表 16.6 Bitmap 类的静态方法

方法名（只列出部分方法）	说　明
createBitmap(Bitmap src)	复制位图
createBitmap(Bitmap src,int x ,int y,int w,int h)	从源位图 src 的指定坐标(x,y)开始，截取宽为 w、高为 h 的部分，用于创建新的位图对象
createScaledBitmap(Bitmap src,int w ,int h,boolean filter)	将源位图 src 缩放成宽为 w、高为 h 的新位图
createBitmap(int w ,int h,Bitmap.Config config)	创建一个宽为 w、高为 h 的新位图（config 为位图的内部配置枚举类）
createBitmap(Bitmap src,int x ,int y,int w,int h,Matrix m,boolean filter)	从源位图 src 的指定坐标(x,y)开始，截取宽为 w、高为 h 的部分，按照 Matrix 变换创建新的位图对象

（2）通过 BitmapFactory 的 static Bitmap decodeXxx()方法也可以构造 Bitmap 类的对象。BitmapFactory 的 static Bitmap decodeXxx()方法如表 16.7 所示。

表 16.7　BitmapFactory 的 static Bitmap decodeXxx()方法

方法名（只列出部分方法）	说　明
decodeByteArray(byte[] data, int offset, int length)	从指定字节数组的 offset 位置开始，将长度为 length 的数据解析成位图
decodeFile(String pathName)	将 pathName 对应的文件解析成的位图对象
decodeFileDescriptor(FileDescriptor fd)	将 FileDescriptor 解析成的位图对象
decodeResource(Resource res,int id)	根据给定的资源 id 解析成位图
decodeStream(InputStream in)	把输入流解析成位图

16.3.3　Android 的自绘控件

自绘控件中所展示的内容都是由用户自己绘制的，所有的绘制操作都是在 onDraw()方法中进行的。自定义控件是 View 类的直接子类，例如，最常使用的 TextView、ImageView 等类就是 View 类的直接子类，它们也可以看成自绘控件。自绘控件的使用步骤如下：

（1）编写自定义属性的 XML 文件。

（2）继承 View 类，覆写构造函数，获取自定义属性。

（3）覆写 onMeasure()方法，测量并设置控件的宽度和高度。

（4）覆写 onDraw()方法，绘制控件。

（5）设置事件。

下面通过自绘控件实现一个简单的计数器，每单击计数器一次，其计数值就加 1 并显示出来。创建继承自 View 类的 CounterView 类，实现 OnClickListener 接口，代码如下：

```
public class CounterView extends View implements OnClickListener {
    //定义画笔
    private Paint mPaint;
    //用于获取文本的宽度和高度
    private Rect mBounds;
    //计数值，每单击一次计数器，其计数值就增加 1
    private int mCount;
    public CounterView(Context context, AttributeSet attrs) {
        super(context, attrs);
        //初始化画笔、Rect
        mPaint = new Paint(Paint.ANTI_ALIAS_FLAG);
        mBounds = new Rect();
        //本控件（计数器）的单击事件
        setOnClickListener(this);
    }
    @Override
    protected void onDraw(Canvas canvas) {
        super.onDraw(canvas);
        mPaint.setColor(Color.BLUE);
```

```
            //绘制一个填充色为蓝色的矩形
            canvas.drawRect(0, 0, getWidth(), getHeight(), mPaint);
            mPaint.setColor(Color.YELLOW);
            mPaint.setTextSize(50);
            String text = String.valueOf(mCount);
            //获取文本的宽度和高度
            mPaint.getTextBounds(text, 0, text.length(), mBounds);
            float textWidth = mBounds.width();
            float textHeight = mBounds.height();
            //绘制字符串
            canvas.drawText(text, getWidth() / 2 - textWidth / 2, getHeight() / 2 + textHeight / 2, mPaint);
        }
        @Override
        public void onClick(View v) {
            mCount ++;
            //重绘
            invalidate();
        }
}
```

定义布局文件，代码如下：

```
<LinearLayout xmlns:android="http://schemas.android.com/apk/res/android"
    android:id="@+id/main_layout"
    android:layout_width="match_parent"
    android:layout_height="match_parent"
    android:orientation="vertical" >
    <com.example.demo.CounterView
        android:id="@+id/counter_view"
        android:layout_width="100dp"
        android:layout_height="100dp"
        android:layout_gravity="center_horizontal|top"
        android:layout_margin="20dp" />
</LinearLayout>
```

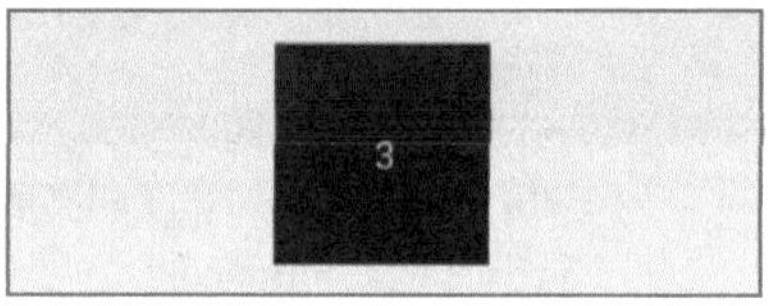

自绘控件运行效果如图 16.3 所示。

图 16.3　自绘控件运行效果

16.4　开发实践：智能医疗仪表图形动态显示

16.4.1　开发设计

本任务采用 Android 中的动态图形绘制类来实现智能医疗仪表图形动态显示。主界面采用线性布局（LinearLayout），里面内嵌一个自定义视图类 MyView 来显示动态图像，背景的网格通过设置背景图片来实现。智能医疗仪表图形动态显示界面如图 16.4 所示，界面布局结构如图 16.5 所示。

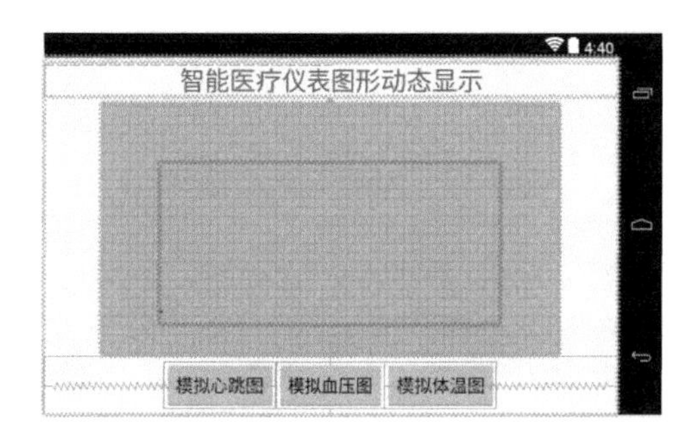

图 16.4　智能医疗仪表图形动态显示界面

图 16.5　界面布局结构

本任务主要编写 MainActivity.java 与 MyView.java 程序文件，以及 activity_main.xml 布局文件。项目目录结构如图 16.6 所示。

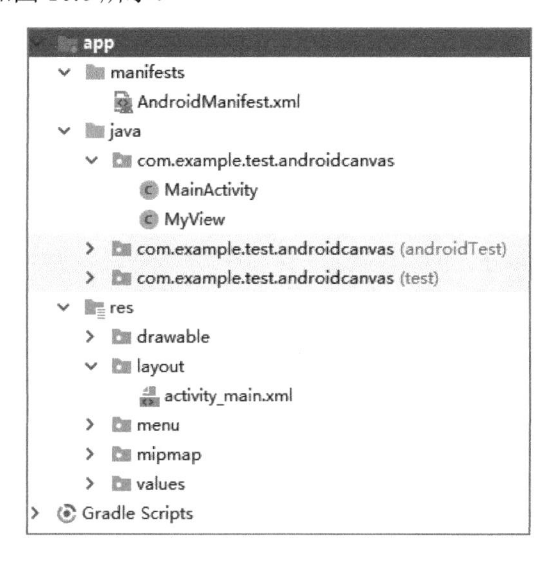

图 16.6　项目目录结构

项目开发步骤如下：

（1）编写 activity_main.xml 布局文件，设置界面显示的控件布局。

（2）编写 MyView.java 程序，MyView 类继承自 View 类，在其 onDraw()方法中实例化一个画笔对象，通过 canvas.drawCircle()方法进行圆点的绘制，在主界面程序中更新坐标。

（3）编写 MainActivity.java 程序，覆写 updatePoint()方法，该方法用于更新要绘制圆点的坐标，坐标是通过随机数产生的。

（4）当 MyHandler 类接收到消息时，对界面进行更新。MyHandler 类处于主线程，可以通过 updatePoint()方法更新坐标、通过 myView.invalidate()方法对界面更新、动态绘制圆点。

（5）在 onCreate()方法中实例化 MyHandler 类，myView 类是主线程里的 UI，不能在子线程里更新，所以要实例化一个匿名线程对象，通过线程向 MyHandler 对象发送消息来更新界面。

16.4.2 功能实现

1. 界面布局设计

```
<?xml version="1.0" encoding="utf-8"?>
<RelativeLayout xmlns:android="http://schemas.android.com/apk/res/android"
    android:layout_width="fill_parent"
    android:layout_height="fill_parent"
    android:orientation="vertical" >
    <TextView
        android:id="@+id/text_header"
        android:layout_width="fill_parent"
        android:layout_height="wrap_content"
        android:layout_marginBottom="5dp"
        android:layout_marginTop="5dp"
        android:gravity="center"
        android:text="@string/app_name"
        android:textSize="26sp" />
    <RelativeLayout
        android:layout_width="fill_parent"
        android:layout_height="fill_parent"
        android:layout_below="@+id/text_header"
        android:layout_marginTop="0dp"
        android:layout_marginRight="60dp"
        android:layout_marginLeft="60dp"
        android:layout_marginBottom="60dp"
        android:id="@+id/bk"
        android:background="@drawable/time_bk">
        <com.x210.draw.MyView
            android:layout_width="fill_parent"
            android:layout_height="fill_parent"
            android:layout_marginTop="60dp"
            android:layout_marginBottom="30dp"
            android:layout_marginLeft="60dp"
            android:layout_marginRight="60dp"
            android:id="@+id/myView"
            />
    </RelativeLayout>
    <RelativeLayout
        android:layout_width="match_parent"
        android:layout_alignParentBottom="true"
        android:layout_height="60dp"
        android:orientation="horizontal">
        <Button
            android:id="@+id/button1"
```

```
            android:layout_toLeftOf="@+id/button2"
            android:layout_alignBottom="@+id/button2"
            android:layout_width="wrap_content"
            android:layout_height="wrap_content"
            android:text="模拟心跳图" />
        <Button
            android:id="@+id/button2"
            android:layout_centerInParent="true"
            android:layout_width="wrap_content"
            android:layout_height="wrap_content"
            android:text="模拟血压图" />
        <Button
            android:id="@+id/button3"
            android:layout_toRightOf="@+id/button2"
            android:layout_alignBottom="@+id/button2"
            android:layout_width="wrap_content"
            android:layout_height="wrap_content"
            android:text="模拟体温图" />
    </RelativeLayout>
</RelativeLayout>
```

2. 主程序设计

```
public class MainActivityDraw extends Activity {
    private MyView myView;
    private MyHandler mHandler;
    private static final int UPDATE = 1;
    @Override
    public void onCreate(Bundle savedInstanceState) {
        super.onCreate(savedInstanceState);
        setContentView(R.layout.main);
        //myView 类是主线程里的 UI，所以不能在子线程里更新
        myView = (MyView) findViewById(R.id.myView);
        mHandler = new MyHandler();//在主线程中创建 Handler 对象，并绑定 Handler 对象与主线程
        //匿名线程对象
        new Thread() {
            public void run() {
                while (!Thread.currentThread().isInterrupted()) {
                    Message msg = mHandler.obtainMessage(UPDATE);
                    mHandler.sendMessage(msg);
                    try {
                        Thread.sleep(10);
                    } catch (InterruptedException e) {
                        e.printStackTrace();
                    }
                }
            }
```

```
        }.start();
    }
    private class MyHandler extends Handler {
        @Override
        public void handleMessage(Message msg) {
            super.handleMessage(msg);
            switch (msg.what) {
                //在收到消息时对界面进行更新，Handler 对象处于主线程，可以对 UI 进行更新
                case UPDATE:
                    updatePoint();
                    myView.invalidate();
                    break;
            }
        }
    }
    //更新坐标
    private void updatePoint() {
        float h = myView.getHeight();
        float w = myView.getWidth();
        float x = myView.getX();
        float y = myView.getY();
        x = x + 1;
        if(x > w) {
            x = 0;
        }
        y = y + getRandomY();
        if (y >= h || y < 0) {
            y = 300;
        }
        myView.setX(x);
        myView.setY(y);                        //很关键，要将更新后的 y 值传递过去
    }
    private int getRandomY() {
        Random r = new Random();
        return r.nextInt(10) - 5;
    }
}
```

3. 动态图像显示设计

```
package com.x210.draw;
import android.content.Context;
import android.graphics.Canvas;
import android.graphics.Color;
import android.graphics.Paint;
import android.util.AttributeSet;
import android.view.View;
```

```
public class MyView extends View {
    //图形当前坐标
    private float x = 0, y = 300;
    public MyView(Context context, AttributeSet attrs) {
        super(context, attrs);
        this.setFocusable(true);
    }
    public float getX() {
        return x;
    }
    public void setX(float x) {                    //形参 x 变量的数据类型由 int 改为 float
        this.x = x;
    }
    public float getY() {
        return y;
    }
    public void setY(float y) {                    //形参 y 变量的数据类型由 int 改为 float
        this.y = y;
    }
    protected void onDraw(Canvas canvas) {
        super.onDraw(canvas);
        //实例化画笔
        Paint p = new Paint();
        //设置画笔颜色
        p.setColor(Color.RED);
        //画图
        canvas.drawCircle(x, y, 10, p);
    }
}
```

16.5 任务验证

在 Android Studio 开发环境中打开本任务的例程，编译通过后运行程序。程序运行效果如图 16.7 所示。

图 16.7 程序运行效果

16.6 开发小结

本任务主要介绍了动态图形的绘制和图形特效的实现。通过本任务的学习，读者可以掌握动态图形绘制的基本思路，以及利用 Matrix 类实现图形旋转、缩放和平移的方法，从而实现智能医疗仪表图形动态显示。

16.7 思考与拓展

（1）请谈谈 Bitmap 类和 BitmapFactory 的关系。

（2）请尝试改变本任务中动态图形的颜色。

任务 17

远程控制服务端的通信

本任务主要介绍 Socket 通信的基础知识、传输模式及编程原理。通过本任务的学习，读者可以掌握 Socket 通信的方法，实现基于 Socket 的远程控制服务端的通信。

17.1 开发场景：如何实现远程控制服务端的通信

建立在无线通信技术基础上的物联网真正实现了万物互联，为智能控制和远程控制提供了技术支撑。在物联网中，要如何实现远程控制服务端的通信呢？

17.2 开发目标

（1）知识目标：熟悉 Socket 的传输模式、编程原理，以及 Socket 通信与 HTTP 通信的区别。

（2）技能目标：掌握 Socket 通信方式。

（3）任务目标：实现基于 Socket 的远程控制服务端的通信。

17.3 原理学习：Socket 通信

17.3.1 Socket 传输模式

Socket 又称为套接字，在程序内部提供了与外界通信的端口（即端口通信）。通过建立 Socket 连接，可以为通信双方的数据提供传输通道。Socket 是一种抽象层，应用程序可以通过 Socket 来发送和接收数据。如果将应用程序添加到网络中，则它就可以与处于同一网络中的其他应用程序进行通信。

根据不同的底层协议，Socket 的实现是多样化的。本任务只介绍基于传输控制协议（Transmission Control Protocol，TCP）和用户数据报协议（User Datagram Protocol，UDP）的通信，在 TCP 和 UDP 中，主要的 Socket 类型为流套接字和数据报套接字。流套接字将 TCP

作为其端对端协议，提供了一个可信赖的字节流服务；数据报套接字使用 UDP，提供数据打包发送服务。

基于 TCP 的 Socket 通信在通信双方建立连接后就可以直接进行数据传输，在连接时可实现信息的主动推送，客户端不需要每次都向服务端发送请求。而基于 UDP 的 Socket 通信则提供无连接的数据报服务，在发送数据报之前不需要建立连接，也不对需要数据报进行检查。

1. TCP 简介

TCP 是一种面向连接的、可靠的传输层通信协议，目的是在跨越多个网络通信时，为两个通信端点之间提供一条通信链路。TCP 具有以下特点：

（1）基于流的方式。

（2）面向连接。

（3）可靠通信方式。

（4）在网络状况不佳的情况下会尽量降低系统由于重传带来的带宽开销。

（5）通信连接维护是面向通信的两个端点的，而不考虑中间网段和节点。

2. UDP 简介

UDP 是 ISO 参考模型中一种无连接的传输层协议，提供面向操作的简单非可靠数据传输服务，UDP 直接工作于 IP 协议的上层，具有以下特点：

（1）不可靠连接，在发送数据时，不知道数据的目的地。

（2）无序发送，如果两个数据同时向目的地发送，它们到达的顺序是无法预测的。

（3）轻量级，无序的数据发送，没有跟踪连接等，仅仅是一个基于 IP 的传输层协议。

（4）无数据校验，单独发送数据报，完整性只有到达时才进行检验。

（5）无堵塞控制，UDP 本身不能避免拥挤，堵塞控制需要在应用程序中实现。

（6）UDP 的头部包含很少的字节，比 TCP 头部的消耗少，传输效率较高。

17.3.2 Socket 编程原理

1. 基于 TCP 的 Socket 通信

服务端先声明一个 ServerSocket 对象并且指定端口号；然后调用 ServerSocket 对象的 accept()方法接收客户端发送的数据，accept()方法在没有数据接收时处于堵塞状态，一旦接收到数据，通过输入流接收数据。

客户端创建一个 Socket 对象，指定目标主机（服务端）的 IP 地址和端口号；然后获取客户端发送数据的输出流；最后将要发送的数据写入输出流即可进行基于 TCP 的 Socket 通信。

2. 基于 UDP 的 Socket 通信

服务端先创建一个 DatagramSocket 对象，并且指点监听的端口；然后创建一个空的 DatagramPacket 对象并指定大小，通过 DatagramPacket 的 receive()方法接收客户端发送的数据，receive()方法与 ServerSocket 对象的 accept()方法类似，在没有数据接收时处于堵塞状态。

客户端也首先创建一个 DatagramSocket 对象，并且指点监听的端口；然后创建一个 InetAddress 对象，这个对象是一个网络地址。接着定义要发送的一个字符串，创建一个 DatagramPacket 对象，并指定要将这个数据报发送到网络的哪个地址以及端口号；最后通过 DatagramSocket 的对象的 send()方法发送数据报。代码如下：

```
String str="hello";
byte[] data=str.getBytes();
DatagramPacket packet=new DatagramPacket( data, data.length, serveraddress, 4567);
socket.send(packet);
```

17.3.3　Socket 编程实例

1．基于 TCP 的 Socket 通信实例

（1）客户端代码如下：

```
//向服务端发送数据
public void TCP_sendMsg(String msg) {
    Socket SerSocket = null;
    OutputStream output = null;
    InputStream input = null;
    try {
        SerSocket = new Socket("192.168.100.5", 7788);
        //第一个参数是目标主机名或目标主机的 IP 地址，第二个参数是目标主机的端口号
        output = SerSocket.getOutputStream();
        output.write(msg.getBytes());                    //把数据写入输出流中
        socket.shutdownOutput();
        //结束数据的写入
        input = SerSocket.getInputStream();
        byte[] bArr = new byte[1024];
        int len = -1;
        sBuf = new StringBuffer();
        while ((len = input.read(bArr)) != -1) {
            sBuf.append(new String(bArr, 0, len, Charset.forName("gbk")));
            //得到返回的数据
        }
        //在主线程中更新 UI
        runOnUiThread(new Runnable() {
            @Override
            public void run() {
                mTextView.setText(sBuf.toString());          //将数据返回到界面显示
            }
        });
    } catch (UnknownHostException e) {
        e.printStackTrace();
    } catch (IOException e) {
        e.printStackTrace();
```

```
        } finally {
            try {
                //注意，输出流不需要关闭，因为它只是在 Socket 中得到输出流对象，并没有创建
                if (SerSocket!= null) {
                    SerSocket.close();                          //释放资源，关闭 Socket
                }
            } catch (IOException e) {
                e.printStackTrace();
            }
        }
}
```

（2）服务端代码如下：

```
public void ReceiveMsg() {
        ServerSocket server = null;
        Socket cSocket = null;
        try {
            server = new ServerSocket(7788);//创建一个 ServerSocket 对象，并让这个对象监听 7788 端口
            //通过 ServerSocket 对象的 accept()方法接收客户端发送的数据，同时创建一个 Socket 对象
            //如果客户端没有发送数据，那么该线程就处于阻塞状态
            while(true){
                cSocket = server.accept();
                System.out.println(cSocket.getInetAddress().getHostName());
                System.out.println(cSocket.getInetAddress().getHostAddress());
                //得到当前发送数据 Socket 对象的主机名和 IP 地址
                InputStream input = cSocket.getInputStream();          //得到该 Socket 对象的输入流
                BufferedInputStream bufInStr = new BufferedInputStream(input);
                byte[] bArr = new byte[1024];
                int len = -1;
                while ((len = bufInStr.read(b)) != -1) {          //从输入流中接收客户端发送的数据
                    System.out.println(new String(b, 0, len,"UTF-8"));
                }
                socket.shutdownInput();                                     //结束接收
                OutputStream outputResult = cSocket.getOutputStream();      //不需要关闭
                outputResult.write("ok,我已经收到！ ".getBytes());
                bufInStr.close();                                           //关闭输入流
                cSocket.close();                               //接收 Socket 的数据后释放资源
                cSocket = null;
            }
        } catch (IOException e) {
            e.printStackTrace();
        }
}
```

基于 TCP 的 Socket 通信实例运行效果如图 17.1 所示。

```
Run Server
"D:\adt-bundle-windows\Android Studio\jre\bin\java" ...

192.168.91.1发送：Hello UDP
```

图 17.1　基于 TCP 的 Socket 通信实例运行效果

2. 基于 UDP 的 Socket 通信实例

（1）客户端代码如下：

```
//发送数据给服务端并接收服务端返回的数据
public void UDP_send(String msg) {
    DatagramSocket socket = null;
    try {
        socket = new DatagramSocket(8880);
        //创建 DatagramSocket 对象并绑定一个本地端口号
        byte[] data = msg.getBytes();                    //把字符串转为字节数组
        InetAddress inetAddress = InetAddress.getByName("192.168.1.100");
        //得到 IP 为 192.168.1.100 的对象
        DatagramPacket pack = new DatagramPacket(data, data.length,inetAddress, 8881);
        socket.send(pack);                               //发送数据报
        //接收服务端返回的数据
        byte[] b = new byte[4*1024];                     //创建数组，用于存放接收到的数据
        DatagramPacket pack2 = new DatagramPacket(b, b.length);
        //定义一个 DatagramPacket 对象用来存储接收到的数据
        socket.receive(pack2);                           //接收数据
        //通过 data.getData()方法得到接收到的数据的字节数组对象
        final String result = new String(pack2.getData(),0,pack2.getLength(), "gbk");
        socket.close();                                  //释放资源
        //在线程中更新 UI
        runOnUiThread(new Runnable() {
            @Override
            public void run() {
                mTextView1.setText(result);
            }
        });
    } catch (SocketException e) {
        e.printStackTrace();
    } catch (UnknownHostException e) {
        e.printStackTrace();
    } catch (IOException e) {
        e.printStackTrace();
    }
}
```

（2）服务端代码如下：

```
public void ReceiveMsg(){
    DatagramSocket socket = null;
    try {
        socket = new DatagramSocket(8881);
        //创建 DatagramSocket 对象并绑定一个本地端口号
        while(true){
            byte[] buf = new byte[4*1024];
            DatagramPacket pack = new DatagramPacket(buf, buf.length);
            //创建一个 DatagramPacket 对象
            socket.receive(pack);
            //读取接收到的数据，如果没有发送数据，则该线程处于阻塞状态
            String str = new String(pack.getData(), 0,pack.getLength(),"UTF-8");
            //字符串输出显示
            String ip = pack.getAddress().getHostAddress();
            //得到发送数据的主机 IP 地址
            System.out.println(ip+"发送:"+str);
            //返回数据给客户端
            InetAddress address = pack.getAddress();
            //得到发送数据的主机网络地址对象
            byte[] data = "已收到！".getBytes();
            DatagramPacket p = new DatagramPacket(data, data.length, address, 8880);
            socket.send(p);
        }
    } catch (SocketException e) {
        e.printStackTrace();
    } catch (IOException e) {
        e.printStackTrace();
    }
}
```

基于 UDP 的 Socket 通信实例运行效果如图 17.2 所示。

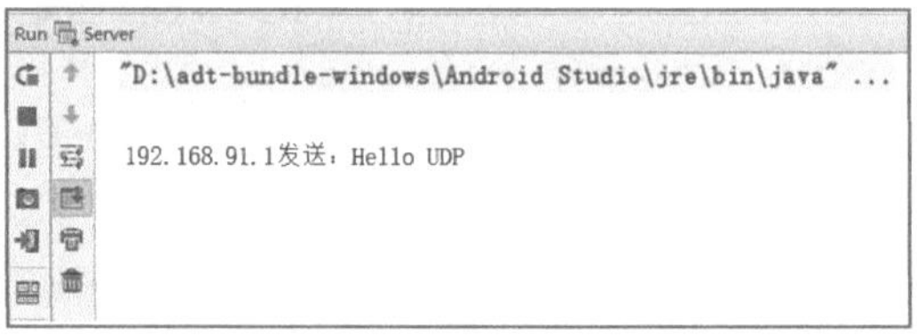

图 17.2　基于 UDP 的 Socket 通信实例运行效果

17.4　开发实践：远程控制服务端通信的实现

17.4.1　开发设计

要实现远程控制服务端的通信，首先需要设计客户端和服务端。客户端采用 Android 应用程序来控制，服务端通过 SocketServer 来接收控制命令。客户端界面如图 17.3 所示，其界面布局结果如图 17.4 所示。

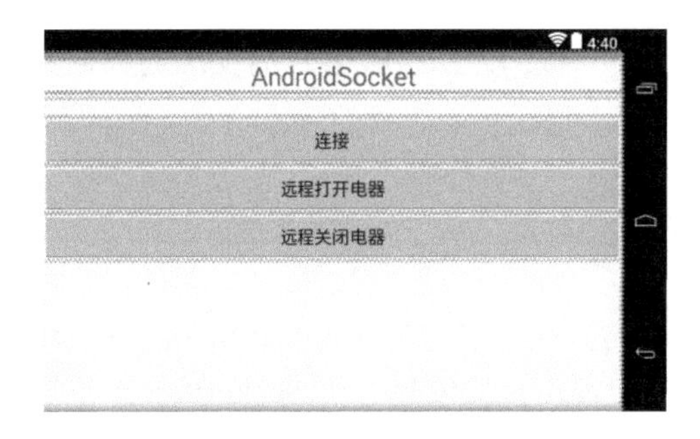

图 17.3　客户端界面

图 17.4　客户端界面布局结构

本项目需要两个软件模块：一个是 app 模块，是 Android 应用；另一个是 SocketServer 模块，是服务端的 Java 程序。在 app 模块中主要编写 MainActivity.java 与 activity_main.xml 布局文件，设置 AndroidManifest.xml 配置文件；在 SocketServer 模块中主要编写 Server.java 程序文件。app 模块的目录结构如图 17.5 所示。

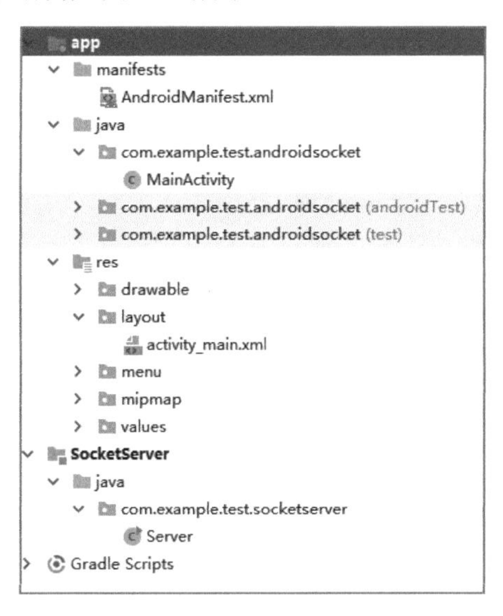

图 17.5　app 模块的目录结构

项目开发步骤如下：

（1）编写 activity_main.xml 布局文件，设置客户端界面显示的控件布局。

（2）设置 AndroidManifest.xml 配置文件，增加应用程序的网络访问权限。

（3）编写 MainActivity.java 程序，首先通过 Socket()实例化一个 clientSocket 对象。

（4）设置“连接”按钮监听器 setOnClickListener，通过 InetSocketAddress 实例化一个 Socket 地址对象，通过 clientSocket.connect(socketAddress)方法连接到服务端。

（5）实例化 clientSocket 的数据输入流、输出流对象。“远程打开电器”按钮和“远程关闭电路”按钮通过 out.writeUTF()方法向服务端发送控制命令。

（6）编写 Server.java 服务端程序，Server 通过一个线程实例对象实现。

（7）在 runServer()方法中主要实现输出服务端的 IP 信息，程序的 main 函数中首先实例化 Server 类，接着调用 runServer()方法。

（8）在覆写的线程 run()方法中，通过 serverSocket.accept()方法响应客户端的连接请求，实例化服务端的数据输入流对象，通过这个对象的 toString()方法接收客户端发送的数据并显示出来。

17.4.2 功能实现

在 Android Studio 开发环境中分别实现服务端和客户端。首先在 Android Studio 开发环境中创建一个项目，然后创建 app 模块，用于开发客户端；接着创建一个 SocketServer 模块，用于开发服务端 Java 程序。

1. 客户端界面布局设计

```
<?xml version="1.0" encoding="utf-8"?>
<LinearLayout xmlns:android="http://schemas.android.com/apk/res/android"
    android:layout_width="fill_parent"
    android:layout_height="fill_parent"
    android:orientation="vertical" >
    <TextView
        android:id="@+id/text_header"
        android:layout_width="fill_parent"
        android:layout_height="wrap_content"
        android:layout_marginBottom="5dp"
        android:layout_marginTop="5dp"
        android:gravity="center"
        android:text="@string/app_name"
        android:textSize="26sp" />
    <TextView
        android:id="@+id/textview"
        android:layout_width="fill_parent"
        android:layout_height="wrap_content"
        android:hint="" />
    <Button
        android:id="@+id/connectbutton"
        android:layout_width="fill_parent"
        android:layout_height="wrap_content"
        android:text="连接" />
    <Button
        android:id="@+id/button1"
        android:layout_width="fill_parent"
        android:layout_height="wrap_content"
        android:text="远程打开电器" />
    <Button
        android:id="@+id/button2"
        android:layout_width="fill_parent"
        android:layout_height="wrap_content"
        android:text="远程关闭电器" />
</LinearLayout>
```

2. 配置文件设计

```
<?xml version="1.0" encoding="utf-8"?>
<manifest xmlns:android="http://schemas.android.com/apk/res/android"
    package="com.x210.sockettest"
    android:versionCode="1"
    android:versionName="1.0">
    <uses-sdk
        android:minSdkVersion="9"
        android:targetSdkVersion="9" />
    <application
        android:allowBackup="true"
        android:icon="@drawable/ic_launcher"
        android:label="@string/app_name"
        android:theme="@style/AppTheme">
        <activity
            android:name="com.x210.sockettest.MyClientActivity"
            android:label="@string/app_name">
            <intent-filter>
                <action android:name="android.intent.action.MAIN" />
                <category android:name="android.intent.category.LAUNCHER" />
            </intent-filter>
        </activity>
    </application>
    <uses-permission android:name="android.permission.INTERNET"></uses-permission>
</manifest>
```

3. 客户端主界面软件设计

```
public class MainActivity extends Activity {
    private TextView mTextView = null;
    private Button connectButton = null;
    private Button onButton = null;
    private Button offButton = null;
    private Socket clientSocket = null;
    private DataOutputStream out = null;
    private int PORT = 9785;
    public MainActivity(){
        clientSocket = new Socket();
    }
    @Override
    public void onCreate(Bundle savedInstanceState) {
        super.onCreate(savedInstanceState);
        //去掉 TitleBar
        requestWindowFeature(Window.FEATURE_NO_TITLE);
        setContentView(R.layout.activity_main);
        //如果一个应用程序在主线程中请求网络操作，将会抛出此异常。
        //解决方案有两个，一是使用 StrictMode，二是使用线程来操作网络请求。
```

```
if (android.os.Build.VERSION.SDK_INT > 9) {
    StrictMode.ThreadPolicy policy = new StrictMode.ThreadPolicy.Builder().permitAll().build();
    StrictMode.setThreadPolicy(policy);
}
mTextView = (TextView) this.findViewById(R.id.textview);
connectButton = (Button) this.findViewById(R.id.connectbutton);
onButton = (Button) this.findViewById(R.id.button1);
onButton.setEnabled(false);
offButton = (Button) this.findViewById(R.id.button2);
offButton.setEnabled(false);
// “连接”按钮监听器
connectButton.setOnClickListener(new View.OnClickListener() {
    @Override
    public void onClick(View v) {
        try{
            if(clientSocket.isConnected()){
                displayToast("已经连接！");
                mTextView.setText("已经连接！");
            }
            else {
                //实验时，查看主机的 IP
                //输入命令：
                InetAddress address = InetAddress.getByName("192.168.100.24");
                SocketAddress socketAddress = new InetSocketAddress(address,PORT);
                clientSocket.connect(socketAddress);//实例化对象并连接到服务端
                if(clientSocket.isConnected()){
                    displayToast("连接成功！");
                    onButton.setEnabled(true);
                    offButton.setEnabled(true);
                    mTextView.setText("连接成功！");
                    new DataInputStream(clientSocket.getInputStream());
                    out = new DataOutputStream(clientSocket.getOutputStream());
                }else{
                    displayToast("连接失败！");
                }
            }
        }
        catch (UnknownHostException e) {
            System.out.println("未知主机地址异常！");
            e.printStackTrace();
        }
        catch (IOException e) {
            System.out.println("clientSocket 输入输出流异常！");
            e.printStackTrace();
        }
    }
});
```

```
        //“远程打开电器”按钮和“远程关闭电器”按钮监听器
        onButton.setOnClickListener(new View.OnClickListener() {
            @Override
            public void onClick(View v) {
                //获得 EditText 的内容
                String text = "发送远程打开电器命令。" + "\r\n";
                try {
                    out.writeUTF(text);
                    displayToast("发送成功！");
                }
                catch (IOException e) {
                    e.printStackTrace();
                }
            }
        });
        offButton.setOnClickListener(new View.OnClickListener() {
            @Override
            public void onClick(View v) {
                //获得 EditText 的内容
                String text = "发送远程关闭电器命令。" + "\r\n";
                try {
                    out.writeUTF(text);
                    displayToast("发送成功！");
                }
                catch (IOException e) {
                    e.printStackTrace();
                }
            }
        });
    }
    //显示 Toast 函数
    private void displayToast(String s) {
        Toast.makeText(this, s, Toast.LENGTH_SHORT).show();
    }
    @Override
    public boolean onCreateOptionsMenu(Menu menu) {
        //Inflate the menu; this adds items to the action bar if it is present.
        getMenuInflater().inflate(R.menu.menu_main, menu);
        return true;
    }
}
```

4．服务端软件设计

```
public class Server implements Runnable {
    private Thread thread = null;
    private ServerSocket serverSocket = null;
```

```
    private Socket socket = null;
    private DataInputStream in = null;
    private boolean status = true;
    private int PORT = 9785;                    //服务端
    Server() {
        thread = new Thread(this);
    }
    public static void main(String[] args) {
        Server server = new Server();
        server.runServer();                     //运行 Server
    }
    @Override
    public void run() {
        try {
            socket = serverSocket.accept();
            if (socket != null) {
                System.out.println("有客户端连接，客户端的地址为：" + socket.getInetAddress());
            }
        } catch (IOException ioe) {
            System.out.println("正在等待客户端！");
            ioe.printStackTrace();
        }
        try {
            in = new DataInputStream(socket.getInputStream());
        } catch (IOException e) {
            System.out.println("Socket 输入输出流异常！");
            e.printStackTrace();
        }
        while (status) {
            try {
                //接收客户端发送的数据
                String text = in.readUTF().toString();
                System.out.println("服务端读取的数据：" + text);
            } catch (IOException e) {
                System.out.println("客户端已经离开");
                System.out.println(e.getMessage());
                e.printStackTrace();
            }
        }
    }
    public void runServer() {
        try {
            serverSocket = new ServerSocket(PORT);
            System.out.println("服务端启动！");
        } catch (IOException ioe) {
            System.out.println("服务端已经启动，ServerSocket 对象不能重复创建！");
            //ioe.printStackTrace();
```

```
            }
            if (serverSocket != null) {
                thread.start();
            }
        }
    }
```

17.5　任务验证

在 Android Studio 开发环境中打开本任务的例程，编译通过后运行程序。

（1）服务端的运行效果如图 17.6 所示。

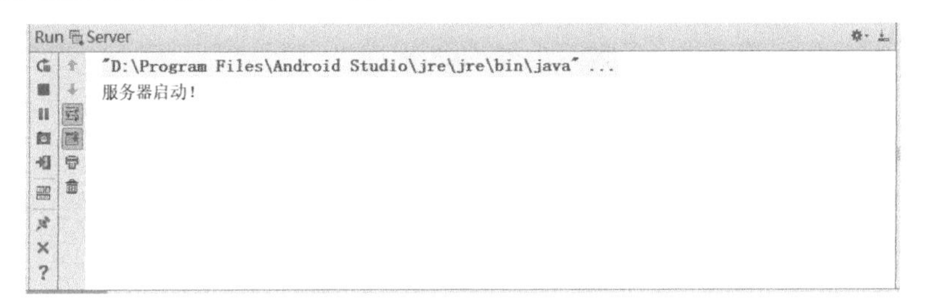

图 17.6　服务端的运行效果

（2）客户端的运行效果如图 17.7 所示。

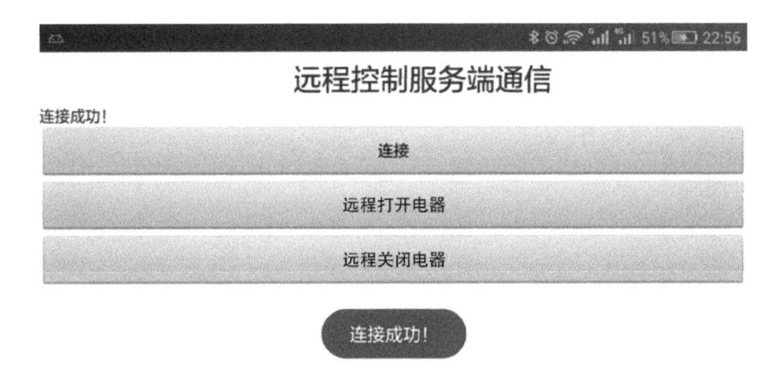

图 17.7　客户端的运行效果

（3）客户端发送远程控制命令后服务端的输出信息如图 17.8 所示。

图 17.8　客户端发送远程控制命令后服务端的输出信息

17.6 开发小结

本任务主要介绍Socket的传输模式和编程原理，给出了基于TCP和基于UDP的Socket通信实例。通过本任务的学习，读者可以理解Socket的传输模式和编程原理，实现基于TCP和基于UDP的Socket通信，实现远程控制服务端通信。

17.7 思考与拓展

（1）请谈谈Socket通信与HTTP通信的区别。

（2）请尝试向服务端发送一条调节电器的命令。

第4篇

物联网 Android 应用开发

本篇以实例的形式讲述物联网 Android 应用开发。本篇共 2 个任务：

任务 18 为物联网系统框架及 Android 开发接口。

任务 19 为仓库环境管理系统的设计。

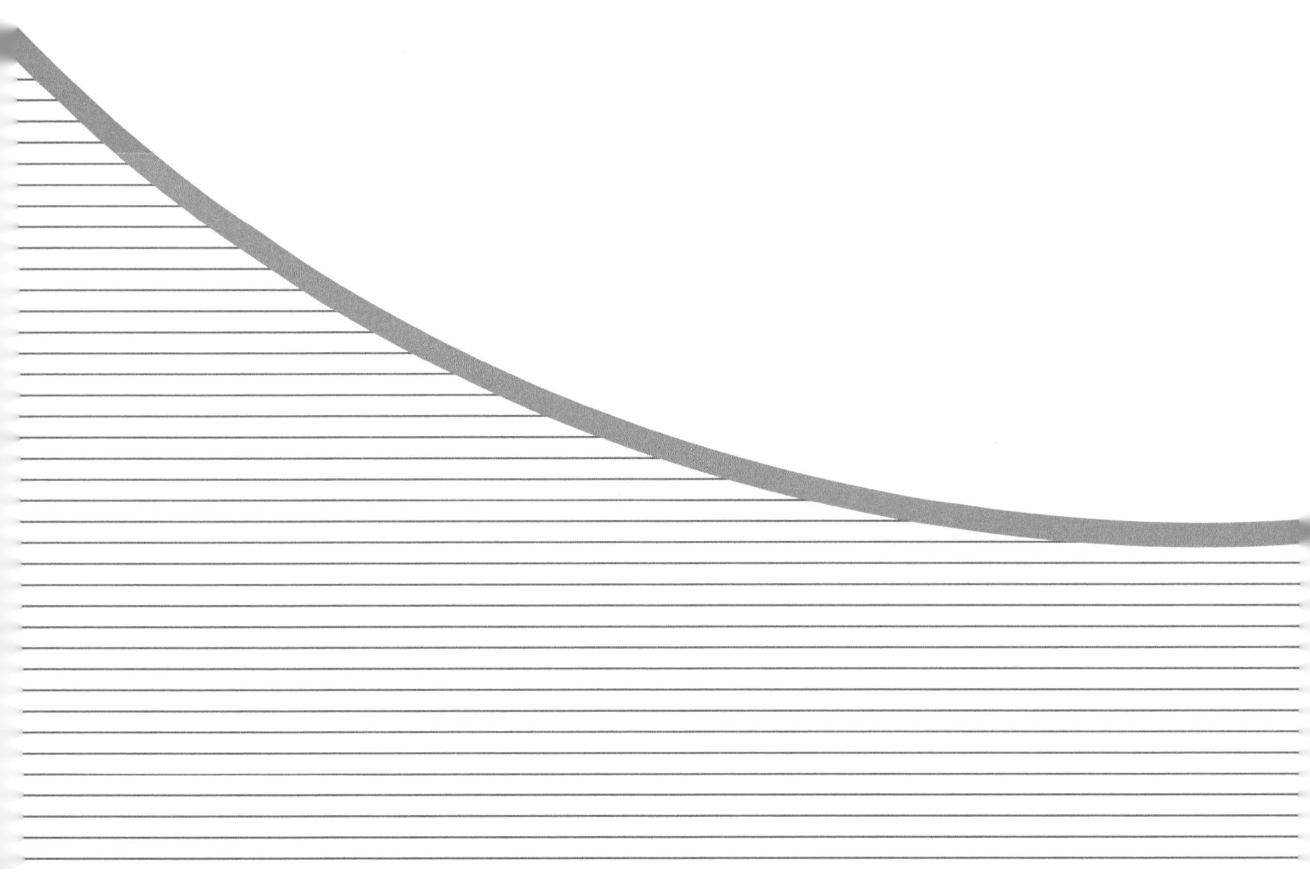

任务 18

物联网系统框架及 Android 开发接口

本任务主要介绍物联网系统框架及智云物联平台的 Android 开发接口，包括实时连接接口、历史数据接口、自动控制接口等，并通过案例来介绍 Android 库 API 编程接口的使用方法。

18.1 开发场景：物联网系统框架

典型的物联网应用需要完成传感器数据的采集、存储、加工和处理等工作，物联网系统框架如图 18.1 所示。

图 18.1　物联网系统框架

18.2 开发目标

（1）知识目标：了解物联网系统框架和 Android 开发接口。

（2）技能目标：理解物联网系统框架、掌握开发调试工具使用，实现基于 Android 接口的实时连接开发。

（3）任务目标：实现基于 Android 接口的实时连接开发。

18.3 原理学习：Android 开发接口

18.3.1 Android 开发接口

智云物联平台提供五大应用接口：实时连接（WSNRTConnect）、历史数据（WSNHistory）、摄像头（WSNCamera）、自动控制（WSNAutoctrl）、用户数据（WSNProperty）。智云物联平台开发接口如图 18.2 所示。

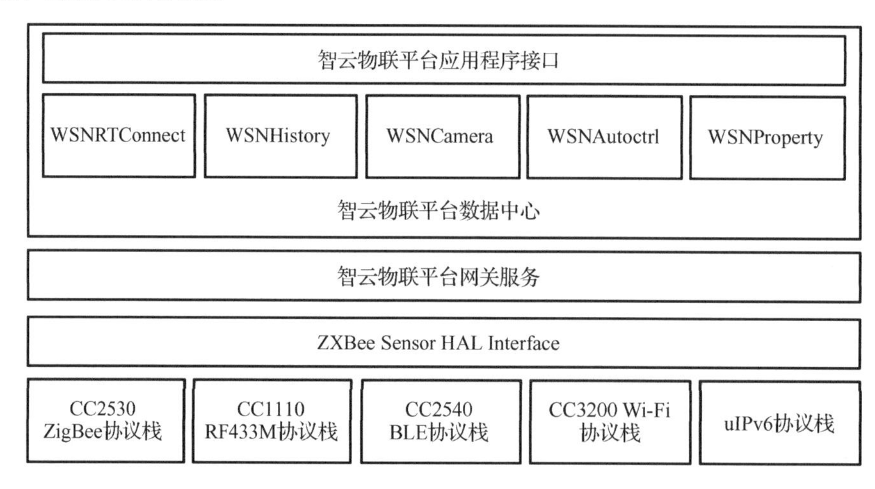

图 18.2 智云物联平台开发接口

针对 Android 应用程序开发，智云物联平台提供了应用接口库 libwsnDroid-20151103.jar，在编写 Android 应用程序时，可以导入该应用接口库，然后调用相应的方法即可。部分 Android 开发接口如下。

1. 实时连接接口

实时连接接口可以推送传感实时数据、执行控制命令、地理位置信息、SMS 短消息等，同时还提供用户信息及通知的统计功能，可方便开发者进行后续开发及运营。实时连接接口如表 18.1 所示。

表 18.1　实时连接接口

函　　数	参 数 说 明	功　　能
new WSNRTConnect(String myZCloudID, String myZCloudKey);	myZCloudID：智云账号。 myZCloudKey：智云密钥	创建实时数据，并初始化智云账号及密钥
connect()	无	建立实时数据推送服务的连接
disconnect()	无	断开实时数据推送服务的连接

setRTConnectListener(){
onConnect()
onConnectLost(Throwable arg0)
onMessageArrive(String mac, byte[] dat)
}
	onConnect：连接成功操作。
	onConnectLost：连接失败操作。
	onMessageArrive：数据接收操作
		设置监听器，接收实时数据推送服务推送的数据

sendMessage(String mac, byte[] dat)	mac：传感器的 MAC 地址。 dat：发送的数据	发送消息
setServerAddr(String sa)	sa：数据中心服务器地址及端口	设置/改变数据中心服务器的地址及端口号
setIdKey(String myZCloudID, String myZCloudKey);	myZCloudID：智云账号。 myZCloudKey：智云密钥	设置/改变智云账号及密钥（需要重新断开连接）

2．历史数据接口

历史数据接口可以完成与云存储服务器的数据连接、数据访问存储、数据使用等。历史数据接口如表 18.2 所示。

表 18.2　历史数据接口

函　　数	参 数 说 明	功　　能
new WSNHistory(String myZCloudID, String myZCloudKey);	myZCloudID：智云账号。 myZCloudKey：智云密钥	初始化历史数据对象，并初始化智云账号及密钥
queryLast1H(String channel);	channel：传感器数据通道	查询最近 1 小时的历史数据
queryLast6H(String channel);	channel：传感器数据通道	查询最近 6 小时的历史数据
queryLast12H(String channel);	channel：传感器数据通道	查询最近 12 小时的历史数据
queryLast1D(String channel);	channel：传感器数据通道	查询最近 1 天的历史数据
queryLast5D(String channel);	channel：传感器数据通道	查询最近 5 天的历史数据
queryLast14D(String channel);	channel：传感器数据通道	查询最近 14 天的历史数据
queryLast1M(String channel);	channel：传感器数据通道	查询最近 1 个月（30 天）的历史数据
queryLast3M(String channel);	channel：传感器数据通道	查询最近 3 个月（90 天）的历史数据
queryLast6M(String channel);	channel：传感器数据通道	查询最近 6 个月（180 天）的历史数据
queryLast1Y(String channel);	channel：传感器数据通道	查询最近 1 年（365 天）的历史数据
query();	无	获取所有通道最后一次数据
query(String channel);	channel：传感器数据通道	获取指定通道最后一次数据

续表

函　数	参数说明	功　能
query(String channel, String start, String end);	channel：传感器数据通道。 start：起始时间。 end：结束时间。 时间为 ISO 8601 格式的日期，例如 2010-05-20T11:00:00Z	通过起止时间查询指定时间段的历史数据（根据时间范围默认选择时间间隔）
query(String channel, String start, String end, String interval);	channel：传感器数据通道。 start：起始时间。 end：结束时间。 interval：采样点的时间间隔，详细见后续说明。 时间为 ISO 8601 格式的日期，例如 2010-05-20T11:00:00Z	通过起止时间查询指定时间段、指定时间间隔的历史数据
setServerAddr(String sa)	sa：数据中心服务器地址及端口	设置/改变数据中心服务器地址及端口号
setIdKey(String myZCloudID, String myZCloudKey);	myZCloudID：智云账号。 myZCloudKey：智云密钥	设置/改变智云账号及密钥

3. 自动控制接口

自动控制接口可以实现数据更新、设备状态查询、定时硬件系统控制、定时发送短消息、根据变量触发某个复杂控制策略、实现系统复杂控制等。自动控制接口的实现步骤如下：

（1）为每个传感器、执行器的关键数据和控制量创建一个变量。

（2）新建基础控制策略，在基础控制策略里可以运用上一步创建的变量。

（3）新建复杂控制策略，复杂控制策略可以使用运算符，可以组合基础控制策略。

自动控制接口如表 18.3 所示。

表 18.3　自动控制接口

函　数	参数说明	功　能
new WSNAutoctrl(String myZCloudID, String myZCloudKey);	myZCloudID：智云账号。 myZCloudKey：智云密钥	初始化自动控制对象，并初始化智云账号及密钥
createTrigger(String name, String type, JSONObject param);	name：触发器名称。 type：触发器类型。 param：触发器内容，JSON 格式，创建成功后返回该触发器 ID（JSON 格式）	创建触发器
createActuator(String name,String type,JSONObject param);	name：执行器名称。 type：执行器类型。 param：执行器内容，JSON 格式，创建成功后返回该执行器 ID（JSON 格式）	创建执行器
createJob(String name, boolean enable, JSONObject param);	name：任务名称。 enable：true（使能任务）、false（禁止任务）。 param：任务内容，JSON 格式，创建成功后返回该任务 ID（JSON 格式）	创建任务

续表

函　数	参 数 说 明	功　能
deleteTrigger(String id);	id：触发器 ID	删除触发器
deleteActuator(String id);	id：执行器 ID	删除执行器
deleteJob(String id);	id：任务 ID	删除任务
setJob(String id,boolean enable);	id：任务 ID。 enable：true（使能任务）、false（禁止任务）	设置任务使能开关
deleteSchedudler(String id);	id：任务记录 ID	删除任务记录
getTrigger();	无	查询当前智云账号下的所有触发器内容

18.3.2　Android 开发接口应用实例

1．实时连接接口应用实例

要实现传感器实时数据的发送，只需要在 Android 项目中调用类 WSNRTConnect 的几个方法即可，具体调用方法及步骤如下：

（1）连接服务器地址。服务器地址及端口默认为“zhiyun360.com:28081”，用户可以调用 setServerAddr(sa)方法进行修改。

```
wRTConnect.setServerAddr(zhiyun360.com:28081);                    //设置服务器地址及端口
```

（2）初始化智云账号及密钥。本实例中在 DemoActivity 中设置智云账号及密钥，并在每个 Activity 中直接使用，后续不再陈述。

```
String myZCloudID = "12345678";                                    //智云账号
String myZCloudKey = "12345678";                                   //智云密钥
wRTConnect = new WSNRTConnect(DemoActivity.myZCloudID,DemoActivity.myZCloudKey);
```

（3）建立数据推送服务的连接。

```
wRTConnect.connect();                                              //调用 connect()方法
```

（4）设置数据推送服务监听器，接收实时推送的消息。

```
wRTConnect.setRTConnectListener(new WSNRTConnectListener() {
    @Override
    public void onConnect() {                                      //连接服务器成功
        //TODO Auto-generated method stub
    }
    @Override
    public void onConnectLost(Throwable arg0) {                    //连接服务器失败
        //TODO Auto-generated method stub
    }
    @Override
```

```
        public void onMessageArrive(String arg0, byte[] arg1) {                //数据到达
            //TODO Auto-generated method stub
        }
});
```

（5）实现消息发送。调用 sendMessage()方法向指定的传感器发送消息。

```
String mac = "00:12:4B:00:03:A7:E1:17";                    //目的地址
String dat = "{OD1=1,D1=?}"                                //数据指令格式
wRTConnect.sendMessage(mac, dat.getBytes());               //发送消息
```

（6）断开数据推送服务。

```
wRTConnect.disconnect();
```

2. 历史数据接口应用实例

同理，要获取传感器的历史数据，只需要在 Java 文件中调用 WSNHistory 类的几个方法即可，具体的调用方法及步骤如下：

（1）实例化历史数据对象。

（2）连接服务器，服务器地址及端口默认为“zhiyun360.com:8080”，用户可以调用 setServerAddr(sa) 方法进行修改。

```
wsnHistory.setServerAddr("zhiyun360.com:8080");
```

（3）初始化智云账号及密钥。

```
id = preferences.getString("id",config.getUserId());
key = preferences.getString("key",config.getUserKey());
wsnHistory.setIdKey(id,key);
```

（4）查询历史数据。以下方法为查询自定义时间段的历史数据，如需要查询其他时间段的历史数据，请参考智云 API 的介绍。

```
String result = wsnHistory.queryLast1H(airPressureMac + "_A0");      //查询最近 1 小时的历史数据
String result = wsnHistory.queryLast1M(airPressureMac + "_A0");      //查询最近 1 个月的历史数据
```

18.4 开发实践：建立服务连接

18.4.1 开发设计

本任务主要通过 Android 应用程序与智云服务器建立实时连接。Android 应用程序的界面框架如图 18.3 所示。

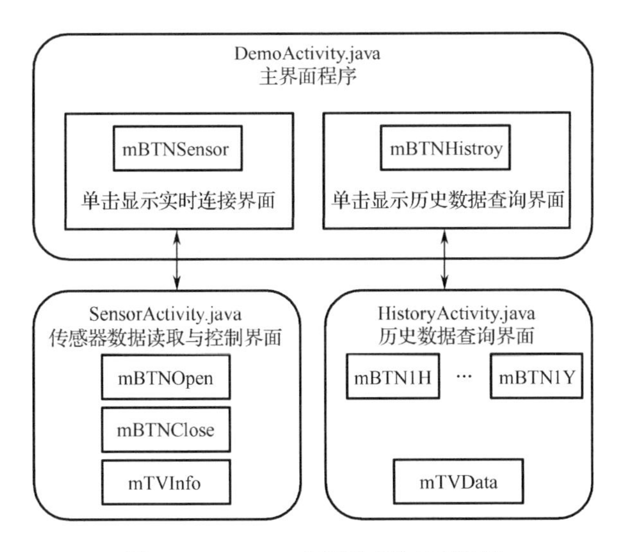

图 18.3　Android 应用程序的界面框架

项目文件说明如表 18.4 所示，项目目录结构如图 18.4 所示。

表 18.4　项目文件说明

文　件	说　明
DemoActivity.java	Android 主界面程序
SensorActivity.java	实时连接接口，实时连接程序
HistoryActivity.java	历史数据接口，历史数据查询程序
AndroidManifest.xml	信息描述文件
main.xml	主界面布局文件
sensor.xml	实时连接布局文件
history.xml	历史数据布局文件

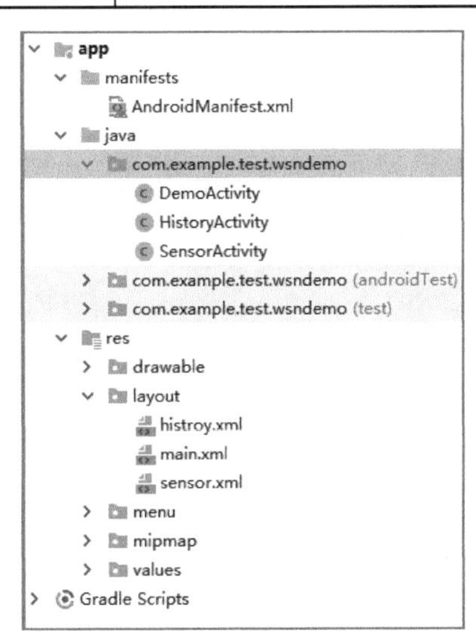

图 18.4　项目目录结构

18.4.2 功能实现

1. 导入应用接口库

在 Android 项目中导入智云物联平台中针对 Android 应用程序开发提供的应用接口库 libwsnDroid-20151103.jar，如图 18.5 所示。

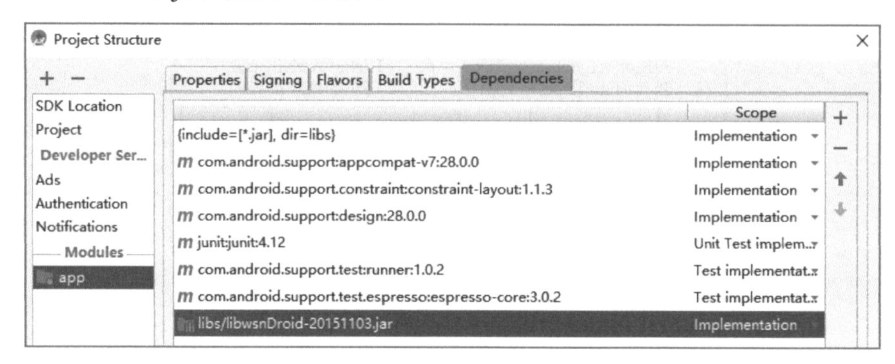

图 18.5 导入 LibwsnDroid-20151103.jar

2. SensorActivity.java 部分代码

```
package com.example.test.wsndemo;
import android.app.Activity;
public class SensorActivity extends Activity  {
    private Button mBTNOpen,mBTNClose;
    private TextView mTVInfo;
    private WSNRTConnect wRTConnect;
    private void textInfo(String s) {
        mTVInfo.setText(mTVInfo.getText().toString() + "\n" + s);
    }
    @Override
    public void onCreate(Bundle savedInstanceState) {
        super.onCreate(savedInstanceState);
        setContentView(R.layout.sensor);
        mBTNOpen = (Button) findViewById(R.id.btnOpen);
        mBTNClose = (Button) findViewById(R.id.btnClose);
        mTVInfo = (TextView) findViewById(R.id.tvInfo);
        wRTConnect = new WSNRTConnect(DemoActivity.myZCloudID,DemoActivity.myZCloudKey);
        wRTConnect.setServerAddr("zhiyun360.com");          //设置智云服务器地址
        wRTConnect.connect();
        mBTNClose.setOnClickListener(new View.OnClickListener() {
            @Override
            public void onClick(View v) {
                //TODO Auto-generated method stub
                String mac = "00:12:4B:00:10:27:A5:19";
                String dat = "{CD1=64,D1=?}";
                textInfo(mac + " <<< " + dat);
```

```
                wRTConnect.sendMessage(mac, dat.getBytes());
            }
        });
        mBTNOpen.setOnClickListener(new OnClickListener() {
            @Override
            public void onClick(View arg0) {
                //TODO Auto-generated method stub
                String mac = "00:12:4B:00:10:27:A5:19";
                String dat = "{OD1=64,D1=?}";
                textInfo(mac + " <<< " + dat);
                wRTConnect.sendMessage(mac, dat.getBytes());
            }
        });
        wRTConnect.setRTConnectListener(new WSNRTConnectListener() {
            @Override
            public void onConnect() {
                //TODO Auto-generated method stub
                textInfo("connected to server");
            }
            @Override
            public void onConnectLost(Throwable arg0) {
                //TODO Auto-generated method stub
                textInfo("connection lost");
            }
            @Override
            public void onMessageArrive(String arg0, byte[] arg1) {
                //TODO Auto-generated method stub
                textInfo(arg0 + " >>> " + new String(arg1));
            }
        });
        textInfo("connecting……");
    }
    @Override
    public void onDestroy() {
        wRTConnect.disconnect();
        super.onDestroy();
    }
}
```

3．HistoryActivity.java 部分代码

```
package com.example.test.wsndemo;
import android.app.Activity;
import java.text.SimpleDateFormat;
import java.util.Date;
public class HistoryActivity extends Activity implements OnClickListener {
    private String channel = "00:12:4B:00:10:27:A5:19_A0";
```

```
private Button mBTN1H, mBTN6H, mBTN12H, mBTN1D, mBTN5D, mBTN14D, mBTN1M,
                     mBTN3M, mBTN6M, mBTN1Y, mBTNSTART, mBTNEND, mBTNQUERY;
private TextView mTVData;
private SimpleDateFormat simpleDateFormat;
private SimpleDateFormat outputDateFormat;
private WSNHistory wHistory;
@Override
public void onCreate(Bundle savedInstanceState) {
    super.onCreate(savedInstanceState);
    setContentView(R.layout.history);
    setTitle("History");
    if (android.os.Build.VERSION.SDK_INT > 9) {
        StrictMode.ThreadPolicy policy = new StrictMode.ThreadPolicy.Builder().permitAll().build();
        StrictMode.setThreadPolicy(policy);
    }
    simpleDateFormat = new SimpleDateFormat("yyyy-M-d");
    outputDateFormat = new SimpleDateFormat("yyyy-MM-dd'T'HH:mm:ss");
    mTVData = (TextView) findViewById(R.id.tvData);
    mBTN1H = (Button) findViewById(R.id.btn1h);
    mBTN6H = (Button) findViewById(R.id.btn6h);
    mBTN12H = (Button) findViewById(R.id.btn12h);
    mBTN1D = (Button) findViewById(R.id.btn1d);
    mBTN5D = (Button) findViewById(R.id.btn5d);
    mBTN14D = (Button) findViewById(R.id.btn14d);
    mBTN1M = (Button) findViewById(R.id.btn1m);
    mBTN3M = (Button) findViewById(R.id.btn3m);
    mBTN6M = (Button) findViewById(R.id.btn6m);
    mBTN1Y = (Button) findViewById(R.id.btn1y);
    mBTNSTART = (Button) findViewById(R.id.btnStart);
    mBTNEND = (Button) findViewById(R.id.btnEnd);
    mBTNQUERY = (Button) findViewById(R.id.query);
    mBTN1H.setOnClickListener(this);
    mBTN6H.setOnClickListener(this);
    mBTN12H.setOnClickListener(this);
    mBTN1D.setOnClickListener(this);
    mBTN5D.setOnClickListener(this);
    mBTN14D.setOnClickListener(this);
    mBTN1M.setOnClickListener(this);
    mBTN3M.setOnClickListener(this);
    mBTN6M.setOnClickListener(this);
    mBTN1Y.setOnClickListener(this);
    mBTNSTART.setOnClickListener(this);
    mBTNEND.setOnClickListener(this);
    mBTNQUERY.setOnClickListener(this);
    wHistory = new WSNHistory(DemoActivity.myZCloudID, DemoActivity.myZCloudKey);
    wHistory.setServerAddr("zhiyun360.com:8080");
}
```

```
@Override
public void onClick(View arg0) {
    //TODO Auto-generated method stub
    mTVData.setText("");
    String result = null;
    try {
        if (arg0 == mBTN1H) {
            result = wHistory.queryLast1H(channel);
        }
        if (arg0 == mBTN6H) {
            result = wHistory.queryLast6H(channel);
        }
        if (arg0 == mBTN12H) {
            result = wHistory.queryLast12H(channel);
        }
        if (arg0 == mBTN1D) {
            result = wHistory.queryLast1D(channel);
        }
        if (arg0 == mBTN5D) {
            result = wHistory.queryLast5D(channel);
        }
        if (arg0 == mBTN14D) {
            result = wHistory.queryLast14D(channel);
        }
        if (arg0 == mBTN1M) {
            result = wHistory.queryLast1M(channel);
        }
        if (arg0 == mBTN3M) {
            result = wHistory.queryLast3M(channel);
        }
        if (arg0 == mBTN6M) {
            result = wHistory.queryLast6M(channel);
        }
        if (arg0 == mBTN1Y) {
            result = wHistory.queryLast1Y(channel);
        }
        if (arg0 == mBTNSTART) {
            new DatePickerDialog(this, new DatePickerDialog.OnDateSetListener() {
                @Override
                public void onDateSet(DatePicker view, int year, int monthOfYear, int dayOfMonth) {
                    mBTNSTART.setText(year + "-" + (monthOfYear + 1) + "-" + dayOfMonth);
                }
            }, 2014, 0, 1).show();
        }
        if (arg0 == mBTNEND) {
            new DatePickerDialog(this, new DatePickerDialog.OnDateSetListener() {
                @Override
```

```
                    public void onDateSet(DatePicker view, int year, int monthOfYear, int dayOfMonth) {
                        mBTNEND.setText(year + "-" + (monthOfYear + 1) + "-" + dayOfMonth);
                    }
                }, 2014, 0, 1).show();
            }
            if (arg0 == mBTNQUERY) {
                Date sdate = simpleDateFormat.parse(mBTNSTART.getText().toString());
                Date edate = simpleDateFormat.parse(mBTNEND.getText().toString());
                String start = outputDateFormat.format(sdate) + "Z";
                String end = outputDateFormat.format(edate) + "Z";
                result = wHistory.query(channel,start, end, "0");
            }
            if(result != null)mTVData.setText(jsonFormatter(result));
        } catch (Exception e) {
            e.printStackTrace();
            Toast.makeText(getApplicationContext(), "查询数据失败，请重试！", Toast.LENGTH_SHORT).
                                                    show();
        }
    }
    public String jsonFormatter(String uglyJSONString) {
        Gson gson = new GsonBuilder().disableHtmlEscaping().setPrettyPrinting().create();
        JsonParser jp = new JsonParser();
        JsonElement je = jp.parse(uglyJSONString);
        String prettyJsonString = gson.toJson(je);
        return prettyJsonString;
    }
}
```

4．DemoActivity.java 部分代码

```
package com.example.test.wsndemo;
import android.app.Activity;
import android.content.Intent;
import android.os.Bundle;
import android.view.View;
import android.view.View.OnClickListener;
import android.widget.Button;
public class DemoActivity extends Activity implements OnClickListener {
    public static String myZCloudID = "123456789";
    public static String myZCloudKey = "ABCDEFGHIJK";
    Button mBTNSensor, mBTNHistory;
    /*Called when the activity is first created.*/
    @Override
    public void onCreate(Bundle savedInstanceState) {
        super.onCreate(savedInstanceState);
        setContentView(R.layout.main);
        setTitle("WSN Demo");
```

```
            mBTNSensor = (Button) findViewById(R.id.btn_sensor);
            mBTNHistory = (Button) findViewById(R.id.btn_history);
            mBTNSensor.setOnClickListener(this);
            mBTNHistory.setOnClickListener(this);
        }
        @Override
        public void onClick(View v) {
            //TODO Auto-generated method stub
            if (v == mBTNSensor) {
                Intent it = new Intent(this,SensorActivity.class);
                startActivity(it);
            }
            if (v == mBTNHistory) {
                Intent it = new Intent(this,HistoryActivity.class);
                startActivity(it);
            }
        }
    }
```

18.5　任务验证

通过实时数据测试工具 ZCloudWebTools 测试与智云服务器的连接，以及实时数据的接收，如图 18.6 所示。

图 18.6　测试与智云服务器的连接，以及实时数据的接收

设置智云账号与智云密钥，代码如下：

```
public class DemoActivity extends Activity implements OnClickListener {
    public static String myZCloudID = "123456789";
    public static String myZCloudKey = "ABCDEFGHIJK";
}
```

硬件设备的 MAC 地址也需要通过 Sensor.java 文件来设置，代码如下：

```
mBTNClose.setOnClickListener(new View.OnClickListener() {
    @Override
    public void onClick(View v) {
        //TODO Auto-generated method stub
        String mac = "00:12:4B:00:10:27:A5:19";
        String dat = "{CD1=64,D1=?}";
        textInfo(mac + " <<< " + dat);
        wRTConnect.sendMessage(mac, dat.getBytes());
    }
});
mBTNOpen.setOnClickListener(new OnClickListener() {
    @Override
    public void onClick(View arg0) {
        //TODO Auto-generated method stub
        String mac = "00:12:4B:00:10:27:A5:19";
        String dat = "{OD1=64,D1=?}";
        textInfo(mac + " <<< " + dat);
        wRTConnect.sendMessage(mac, dat.getBytes());
    }
});
```

历史数据查询功能需要在 HistoryActivity.java 文件中设置通道，代码如下：

```
public class HistoryActivity extends Activity implements OnClickListener {
    private String channel = "00:12:4B:00:10:27:A5:19_A0";
}
```

主界面如图 18.7 所示。

图 18.7　主界面

传感器数据读取与控制界面如图 18.8 所示。

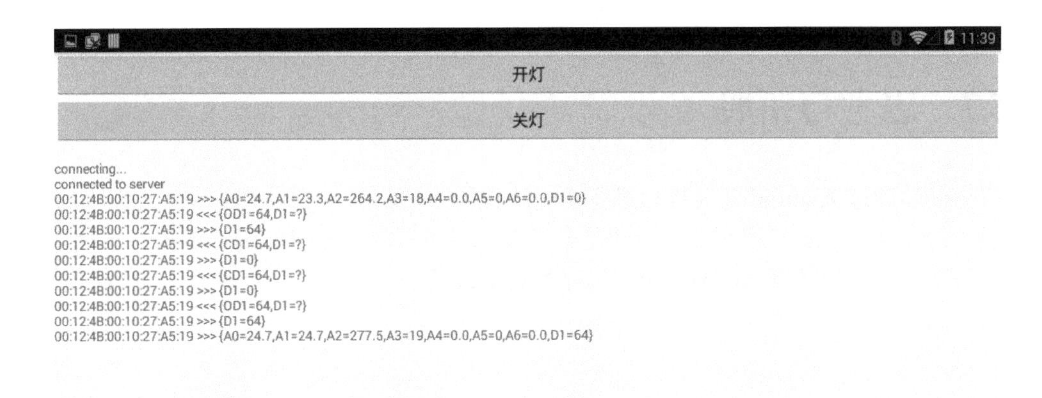

图 18.8　传感器数据读取与控制界面

历史数据查询界面如图 18.9 所示。

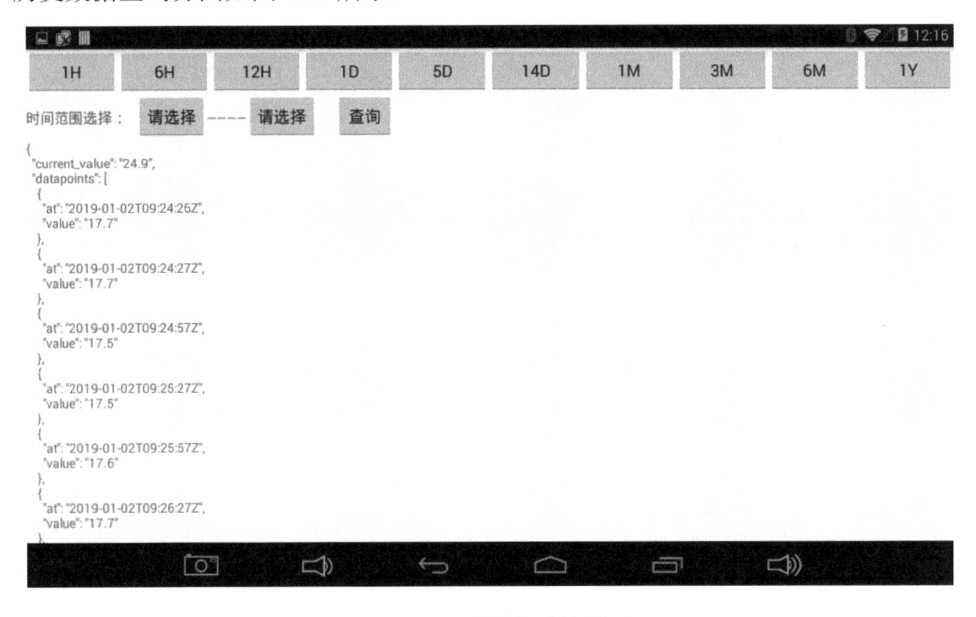

图 18.9　历史数据查询界面

18.6　开发小结

本任务主要介绍物联网系统框架和 Android 开发接口，以及部分 Android 开发接口的使用方法，完成了与智云服务器的连接。

18.7 思考与拓展

（1）简述通过 Android 开发接口进行实时连接的步骤。

（2）假如 Android 应用程序显示的是某个节点设备采集的温度和湿度数据，请描述温度和湿度数据从节点设备到 Android 应用程序的传输过程。

任务19

仓库环境管理系统的设计

本任务主要介绍 Android 应用实例开发，实现实时连接与历史数据接口二次开发，完成仓库环境管理系统的设计。

19.1 开发场景：如何设计仓库环境管理系统

为了保持仓库内温度和湿度的稳定，需要开发仓库环境管理系统来实时监测仓库内的温度和湿度，仓库环境管理系统根据监测到的温度和湿度数据以及设置的阈值来控制风机、空调等设备的开启和关闭，实现仓库环境的无人化管理。

仓库环境管理系统的“运营首页”如图 19.1 所示。

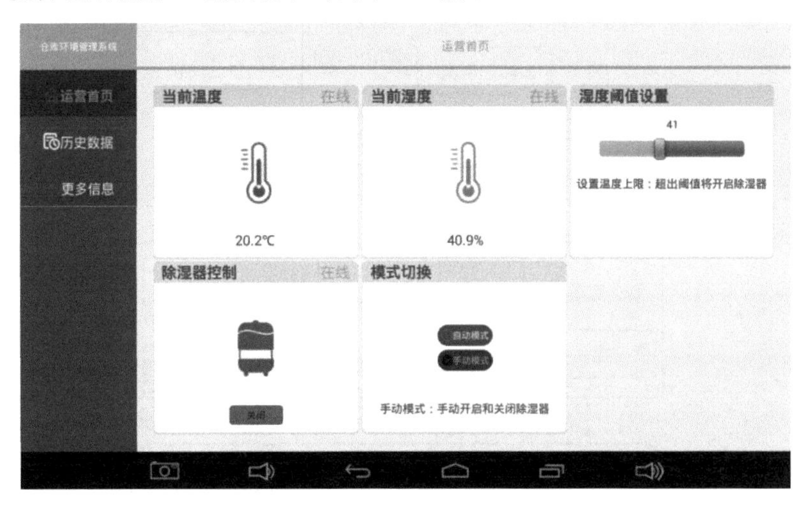

图 19.1　仓库环境管理系统的“运营首页”

19.2 开发目标

（1）知识目标：了解物联网系统的分析方法和 Android 应用程序设计。

（2）技能目标：掌握物联网系统的分析方法，以及 Android 应用程序常用接口的编程方法。

（3）任务目标：通过仓库环境管理系统的设计与实现，实现基于实时连接接口与历史数据接口的二次开发。

19.3 原理学习：仓库环境管理系统分析和Android应用程序设计

19.3.1 仓库环境管理系统分析

1. 系统总体架构设计

仓库环境管理系统基于智云物联平台进行设计，下面根据物联网系统的四层架构模型进行说明。

感知层：通过采集类和控制类传感器实现，温湿度传感器和继电器由 CC2530 控制。

网络层：感知层同 Android 网关之间的无线通信通过 ZigBee 网络实现，Android 网关同智云服务器、上层应用设备间通过局域网进行数据传输。

平台层：主要使用智云物联平台提供的数据存储、交换、分析功能，平台层提供了物联网设备之间的数据存储、访问和控制。

应用层：应用层主要是物联网系统的人机交互接口，通过 PC 端和移动端实现界面友好、操作交互性强的应用。

仓库环境管理系统的硬件总体架构如图 19.2 所示。

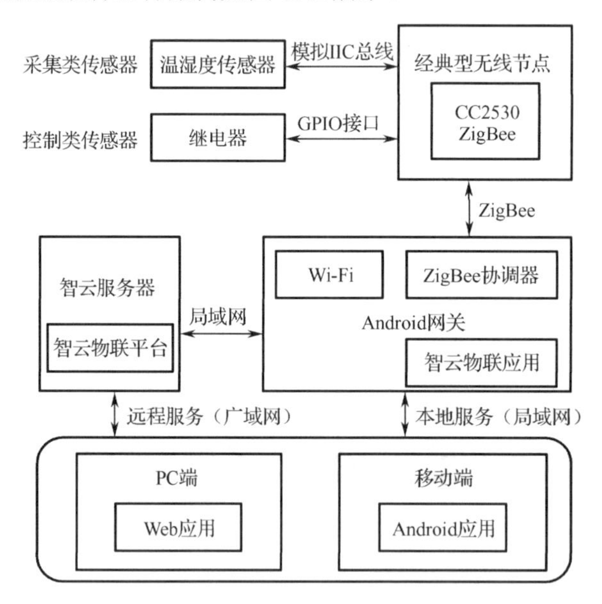

图 19.2 仓库环境管理系统的硬件总体架构

2. 系统功能设计说明

仓库环境管理系统功能可分为两个模块：设备采集控制模块和系统设置模块，如图 19.3

所示。设备采集控制模块包括温湿度传感器的数据采集和继电器的控制；系统设置模块包括设置 ID 和 KEY、设置 MAC 地址、系统版本管理。

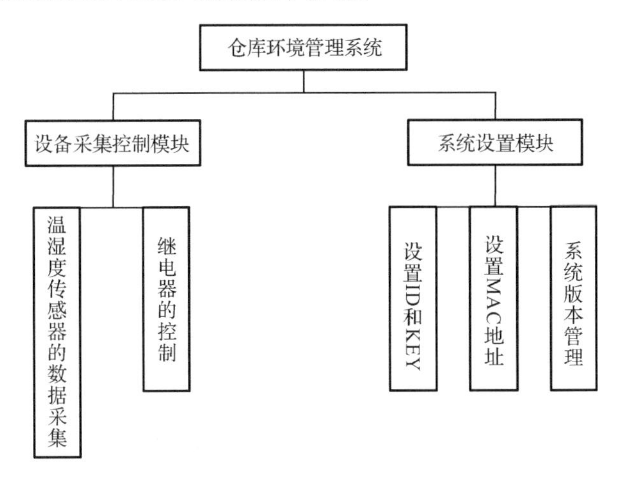

图 19.3　仓库环境管理系统功能的模块划分

仓库环境管理系统能够将温湿度传感器采集的数据发送到智云服务器，智云服务器通过比较采集的数据和设定的阈值，可以控制远程设备（如除湿机）的开关，从而实现仓库环境的无人化管理。

仓库环境管理系统功能需求表设计如表 19.1 所示。

表 19.1　仓库环境管理系统功能需求表

功 能 需 求	说　明
采集数据显示	在上层应用界面实时更新、显示温湿度传感器状态
远程设备的实时控制	通过上层应用程序，对远程设备进行操作
连接智云服务器	智云服务器的参数设置，MAC 地址的设置

3. 系统通信流程分析

仓库环境管理系统通信涉及传感器节点、Android 网关、客户端（Android 和 Web）三个部分，通信流程如图 19.4 所示。

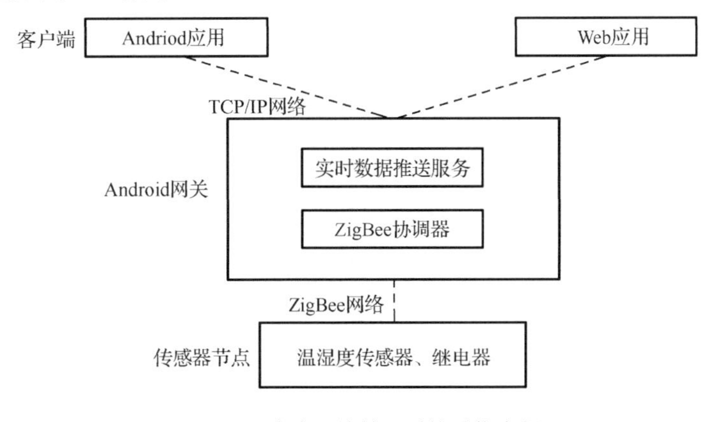

图 19.4　仓库环境管理系统通信流程

4. 硬件选型与原理说明

仓库环境管理系统的硬件主要由 xLab 开发平台的 Android 网关、经典型无线节点（ZXBeeLiteB）、采集类开发平台（Sensor-A）、控制类开发平台（Sensor-B）组成。Android 网关对传感器节点数据进行汇集，经典型无线节点以无线通信方式向 Android 网关发送传感器节点的数据，接收 Android 网关发送的相关控制命令，采集类传感器、控制类传感器连接到经典型无线节点（ZigBee 无线节点），由 CC2530 对相关设备进行采集和控制。

（1）Android 网关。仓库环境管理系统通过 Android 网关进行整个网络的数据汇集，Android 网关运行智云服务程序，连接传感器网络与局域网，同时将数据推送给智云服务器。

Android 网关如图 19.5 所示。

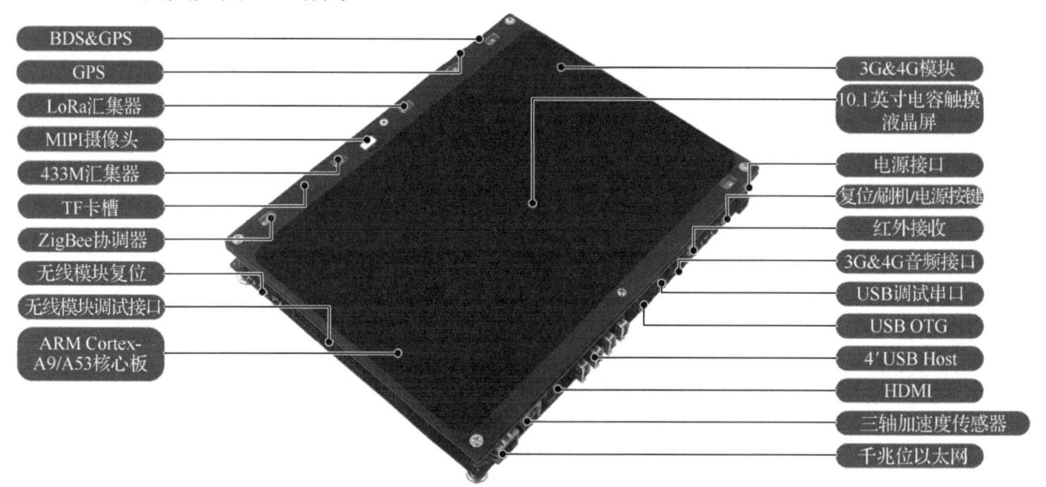

图 19.5　Android 网关

（2）采集类开发平台（Sensor-A）。采集类开发平台包括温湿度传感器、光照度传感器、空气质量传感器、气压海拔传感器、三轴加速度传感器、距离传感器、继电器接口、语音识别传感器等，如图 19.6 所示。

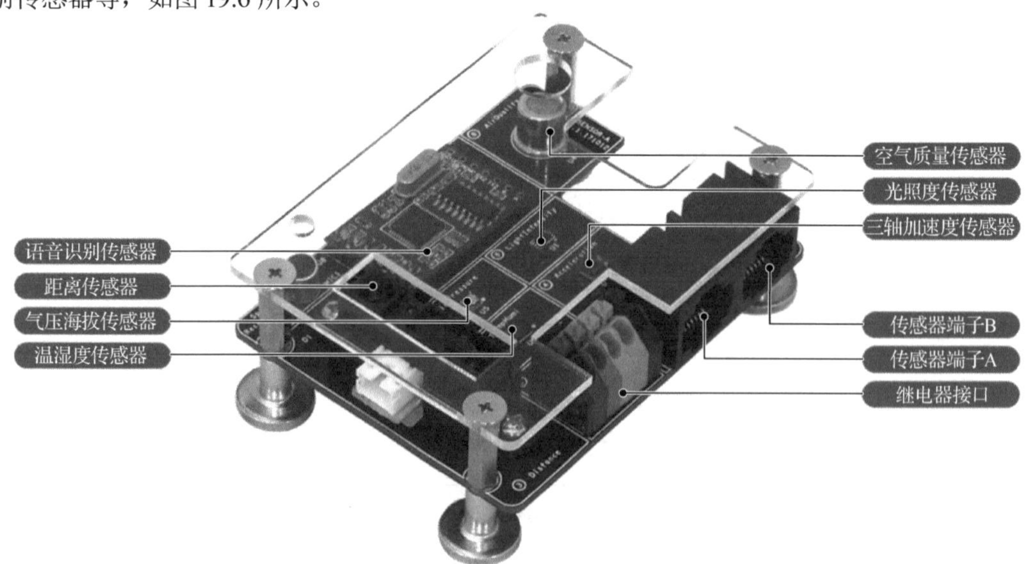

图 19.6　采集类开发平台

（3）控制类开发平台（Sensor-B）。控制类开发平台包括风扇、步进电机、蜂鸣器、LED、RGB 灯、继电器接口等，如图 19.7 所示。

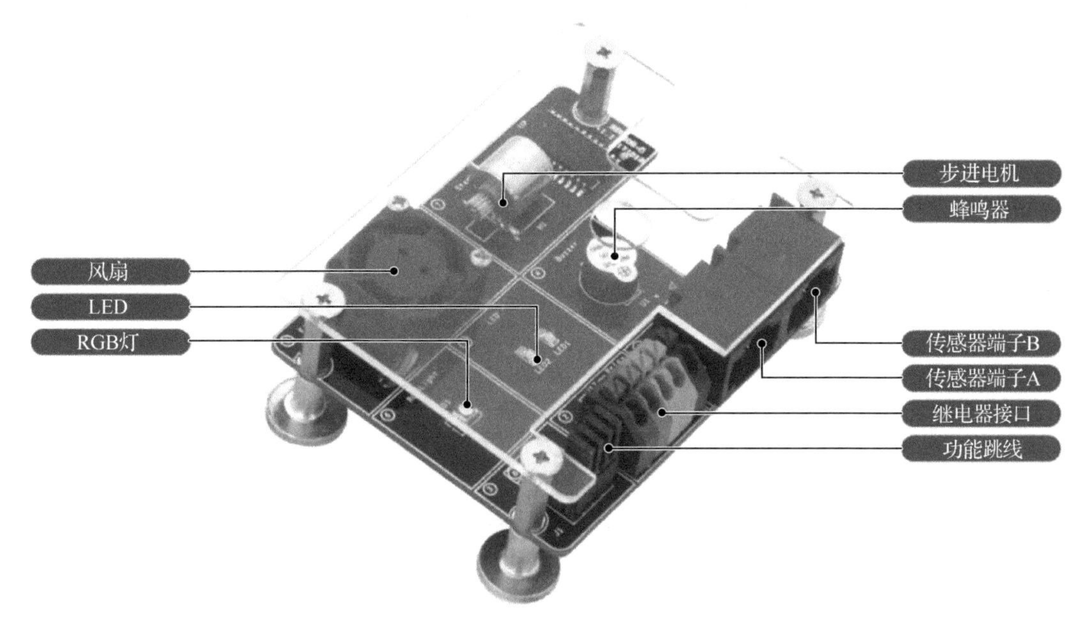

图 19.7　控制类开发平台

19.3.2　Android 应用程序设计

1. 程序框架分析

根据 Android 开发接口的定义，仓库环境管理系统的应用设计主要采用实时数据 API 接口。Android 应用程序框架如图 19.8 所示。

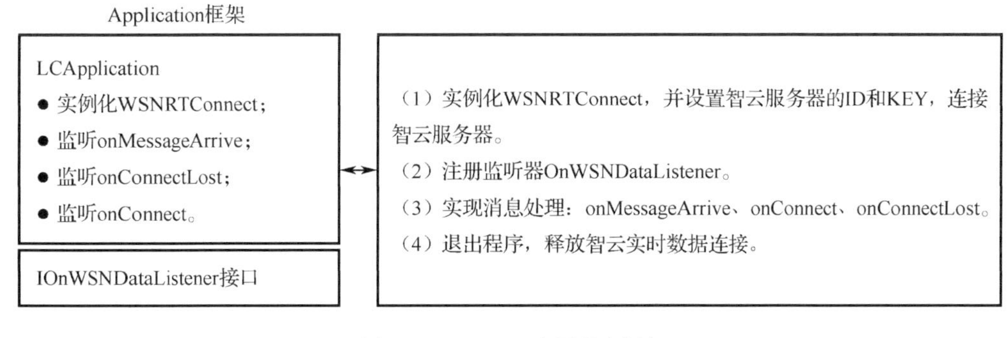

图 19.8　Android 应用程序框架

2. Android 项目框架

仓库环境管理系统的目录结构如图 19.9 所示。

仓库环境管理系统架构说明如表 19.2 所示。

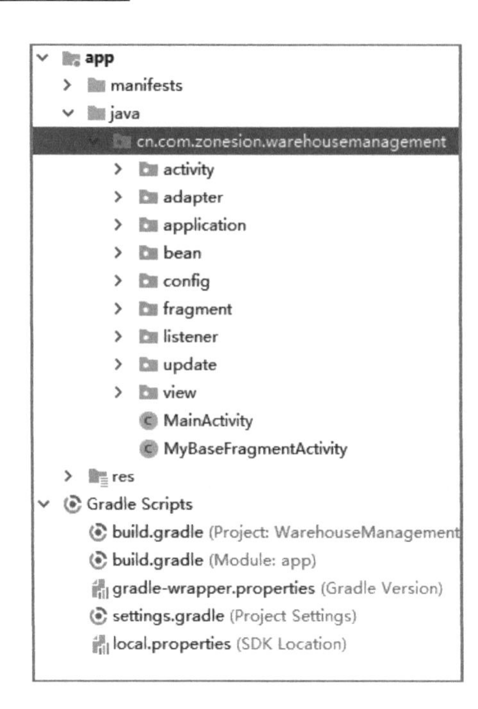

图 19.9 仓库环境管理系统的目录结构

表 19.2 仓库环境管理系统架构说明

类 名	说 明
activity	
IdKeyShareActivity.java	在 IDKey 页面单击“分享”按钮时，会弹出分享 Activity 的二维码图片
adapter	
HdArrayAdapter.java	历史数据显示适配器
application	
LCApplication.java	LCApplication 继承自 Application 类，使用单例模式创建 WSNRTConnect 对象
bean	
HistoricalData.java	历史数据的 Bean 类，将从智云服务器获得的历史数据（JSON 形式）转换成该类对象
IdKeyBean.java	IdKeyBean 类用来描述用户设备的 ID、KEY 和使用的智云服务器的地址 SERVER
config	
Config.java	该类用于对用户设备的 ID、KEY，使用的智云服务器地址及 MAC 地址进行修改
Fragment	
BaseFragment.java	界面基础 Fragment 类
HDFragment.java	湿度历史数据页面
HistoricalDataFragment.java	历史数据显示页面
HomepageFragment.java	首页的 Fragment
IDKeyFragment.java	选择 IDKey 选项时所显示的页面
MacSettingFragment.java	用户设置被检测项的 MAC 地址时所显示的页面

续表

类　名	说　明
MoreInformationFragment.java	更多信息显示页面
RunHomePageFragment.java	运营首页显示页面
temperatureHDFragment.java	温度历史数据显示页面
VersionInformationFragment.java	显示版本等相关信息的页面
listener	
IOnWSNDataListener.java	传感器数据监听器接口
update	
UpdateService.java	应用下载服务类
view	
APKVersionCodeUtils.java	获取当前本地 apk 的版本
CustomRadioButton.java	自定义按钮类
PagerSlidingTabStrip.java	自定义滑动控件类
MainActivity：主界面类	
MyBaseFragmentActivity：系统 Fragment 通信类	

3．Android 接口分析

（1）Application 框架：在 LCApplication.java 类中实例化 WSNRTConnect 类对象，并实现对实时连接的监听。

```
public class LCApplication extends Application implements WSNRTConnectListener{
    private WSNRTConnect wsnrtConnect;                    //创建 WSNRTConnect 对象
    private ArrayList<IOnWSNDataListener> mIOnWSNDataListeners = new ArrayList<>();
    //传感器数据监听器
    private boolean isDisconnected = true;
    //判断当前连接是否断开，true 表示连接断开，false 表示已经连接
    public boolean getIsDisconnected() {                  //属性 isDisconnected 的 getter()方法
        return isDisconnected;
    }
    public void setDisconnected(boolean disconnected) {
        //属性 isDisconnected 的 setter()方法
        isDisconnected = disconnected;
    }
    public WSNRTConnect getWSNRConnect() {
    }
    //注册传感器数据监听器
    public void registerOnWSNDataListener(IOnWSNDataListener li) {
    }
    //取消注册传感器数据监听器
    public void unregisterOnWSNDataListener(IOnWSNDataListener li) {
    }
    @Override
```

```
        public void onCreate() {
        }
        @Override
        public void onConnectLost(Throwable throwable) {
        }
        @Override
        public void onConnect() {
        }
        //消息到达时会自动调用该方法
        @Override
        public void onMessageArrive(String mac, byte[] data) {
        }
}
```

在 onCreate()方法中调用 getWSNRConnect()方法来初始化 WSNRTConnect 对象。

```
public WSNRTConnect getWSNRConnect() {
    if (wsnrtConnect == null) {
        wsnrtConnect = new get WSNRTConnect();              //初始化 WSNRTConnect 对象
    }
    return wsnrtConnect;
}
public void onCreate() {
    super.onCreate();
    wsnrtConnect = getWSNRConnect();
    wsnrtConnect.setRTConnectListener(this);
}
```

其中，当获取到传感器上传的数据时，会调用 onMessageArrive()方法。若要在 Activity 中实现对传感器数据的获取，则需要在 onMessageArrive()方法中分发消息。为了实现消息的分发，定义了一个接口 IOnWSNDataListener，在接口中定义了 void onMessageArrive(String mac, String tag, String val)方法，相关代码如下：

```
//传感器数据监听器
public interface IOnWSNDataListener {
    //发送数据
    void onMessageArrive(String mac, String tag, String val);
    //连接失败
    void onConnectLost();
    //连接成功
    void onConnect();
}
```

定义好接口后，在 LCApplication 中定义传感器数据监听器以及注册传感器监听的方法 registerOnWSNDataListener(IOnWSNDataListener li)。当 Activity 需要获取传感器数据时，需要调用 LCApplication 类中该方法将其加入传感器数据监听器中，相关代码如下：

```
private ArrayList<IOnWSNDataListener> mIOnWSNDataListeners = new ArrayList<>();
//传感器数据监听器
//注册传感器数据监听器
public void registerOnWSNDataListener(IOnWSNDataListener li) {
```

```
        mIOnWSNDataListeners.add(li);
    }
    //取消注册传感器数据监听器
    public void unregisterOnWSNDataListener(IOnWSNDataListener li) {
        mIOnWSNDataListeners.remove(li);
    }
```

在 onMessageArrive()方法中先对传感器上传的数据进行解析，然后将解析得到的数据分发给传感器数据监听器，让其调用接口中的 onMessageArrive()方法。相关代码如下：

```
    public void onMessageArrive(String mac, byte[] data) {
        if (data[0] == '{' && data[data.length - 1] == '}') {
            String sData = new String(data, 1, data.length - 2);
            String[] pDatas = sData.split(",");
            for (String pData : pDatas) {
                String[] tagVal = pData.split("=");
                if (tagVal.length == 2) {
                    for (IOnWSNDataListener li : mIOnWSNDataListeners) {
                        /*实现 IOnWSNDataListener 传感器数据监听接口的类都会自动调用
onMessageArrive()方法*/
                        li.onMessageArrive(mac, tagVal[0], tagVal[1]);
                    }
                }
            }
        }
    }
```

以上为传感器数据处理过程，所有需要获取传感器数据的 Activity，只需要调用 LCApplication 类的 registerOnWSNDataListener(IOnWSNDataListener li)方法，并覆写数据处理函数 onMessageArrive(String mac, String tag, String val)即可。

19.4　开发实践：仓库环境管理系统的设计

19.4.1　开发设计

仓库环境管理系统基于智云物联平台进行设计，本任务主要关注应用层设计。仓库环境管理系统工作流程如图 19.10 所示。

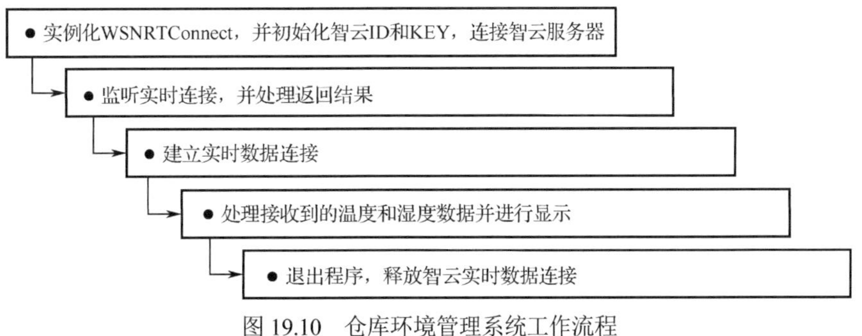

图 19.10　仓库环境管理系统工作流程

19.4.2 功能实现

1. 软硬件设计

（1）应用程序模块。从 LCApplication 获取 WSNRTConnect，建立实时数据连接，注册数据监听器 registerOnWSNDataListener，覆写 onMessageArrive()方法来处理接收到数据，通过按钮单击事件发送传感器控制命令。LCApplication 代码如下：

```
package cn.com.zonesion.warehousemanagement.application;
import java.util.ArrayList;
import cn.com.zonesion.warehousemanagement.listener.IOnWSNDataListener;
import com.zhiyun360.wsn.droid.WSNRTConnect;
import com.zhiyun360.wsn.droid.WSNRTConnectListener;
import android.app.Application;
import android.widget.Toast;
/*LCApplication 继承自 Application 类，在该类中使用单例模式创建 WSNRTConnect 对象，使
WSNRTConnect 对象在整个 App 中是唯一的*/
public class LCApplication extends Application implements WSNRTConnectListener{
    private WSNRTConnect wsnrtConnect;                    //创建 WSNRTConnect 对象
    //传感器数据监听器
    private ArrayList<IOnWSNDataListener> mIOnWSNDataListeners = new ArrayList<>();
    //判断当前连接是否断开，true 表示连接断开，false 表示已经连接
    private boolean isDisconnected = true;
    public boolean getIsDisconnected() {                  //属性 isDisconnected 的 getter()方法
        return isDisconnected;
    }
    public void setDisconnected(boolean disconnected) {
        //属性 isDisconnected 的 setter()方法
        isDisconnected = disconnected;
    }
    public WSNRTConnect getWSNRConnect() {
        if (wsnrtConnect == null) {
            wsnrtConnect = new WSNRTConnect(); //初始化 WSNRTConnect 对象
        }
        return wsnrtConnect;
    }
    //注册传感器数据监听器
    public void registerOnWSNDataListener(IOnWSNDataListener li) {
        mIOnWSNDataListeners.add(li);
    }
    //取消注册传感器数据监听器
    public void unregisterOnWSNDataListener(IOnWSNDataListener li) {
        mIOnWSNDataListeners.remove(li);
    }
    @Override
```

```
        public void onCreate() {
            super.onCreate();
            wsnrtConnect = getWSNRConnect();
            wsnrtConnect.setRTConnectListener(this);
        }
        @Override
        public void onConnectLost(Throwable throwable) {
            Toast.makeText(this, "数据服务断开连接！ ", Toast.LENGTH_SHORT).show();
            for (IOnWSNDataListener li : mIOnWSNDataListeners) {
                //实现了 IOnWSNDataListener 的类都会自动调用 onConnectLost()方法
                li.onConnectLost();
            }
        }
        @Override
        public void onConnect() {
            Toast.makeText(this, "数据服务连接成功！ ", Toast.LENGTH_SHORT).show();
            for (IOnWSNDataListener li : mIOnWSNDataListeners) {
                //实现了 IOnWSNDataListener 的类都会自动调用 onConnect()方法
                li.onConnect();
            }
        }
        //消息到达时会自动调用该方法
        @Override
        public void onMessageArrive(String mac, byte[] data) {
            if (data[0] == '{' && data[data.length - 1] == '}') {
                String sData = new String(data, 1, data.length - 2);
                String[] pDatas = sData.split(",");
                for (String pData : pDatas) {
                    String[] tagVal = pData.split("=");
                    if (tagVal.length == 2) {
                        for (IOnWSNDataListener li : mIOnWSNDataListeners) {
                            //实现 IOnWSNDataListener 的类都会自动调用 onMessageArrive()方法
                            li.onMessageArrive(mac, tagVal[0], tagVal[1]);
                        }
                    }
                }
            }
        }
    }
```

（2）湿度历史数据模块部分代码。

```
package cn.com.zonesion.warehousemanagement.fragment;
import android.content.Context;
import android.content.SharedPreferences;
import android.graphics.Color;
import android.graphics.Typeface;
public class temperatureHDFragment extends BaseFragment implements IOnWSNDataListener{
```

```
    private Spinner airHumiditySearch;
    private LineChartView lineChartView;
    private TextView textDataTip;
    private String airHumidityMac;
    private WSNHistory wsnHistory;
    private LCApplication lcApplication;
    private int position;
    private boolean hasRunned = false;
    private SharedPreferences preferences;
    private Config config;
    private WSNRTConnect wsnrtConnect;

    @Override
    public View initView() {
        View view = View.inflate(mContext, R.layout.temperature_hd_layout, null);
        airHumiditySearch = (Spinner) view.findViewById(R.id.air_humidity_search);
        lineChartView = (LineChartView) view.findViewById(R.id.chart);
        textDataTip = (TextView) view.findViewById(R.id.text_data_tip);
        return view;
    }
    @Override
    public void initData() {
        super.initData();
        initViewAndBindEvent();
    }
     private void initViewAndBindEvent() {
         config = Config.getConfig();
         lcApplication = (LCApplication) getActivity().getApplication();
         lcApplication.registerOnWSNDataListener(this);
         wsnrtConnect = lcApplication.getWSNRConnect();
         preferences = getActivity().getSharedPreferences("user_info", Context.MODE_PRIVATE);
         wsnHistory = new WSNHistory();      //声明一个 WSNHistory 对象
         initSetting();
        //声明一个 ArrayAdapter 对象，用来对 Spinner 进行适配
        ArrayAdapter<String>  adapter  =  new  ArrayAdapter<>(getActivity(),  android.R.layout.simple_
list_item_1, getResources().getStringArray(R.array.airTemperatureSearch));
        //通过 setAdapter()方法将适配器设置给 Spinner
        airHumiditySearch.setAdapter(adapter);
        //绑定 Spinner 的响应事件
        airHumiditySearch.setOnItemSelectedListener(new AdapterView.OnItemSelectedListener() {
            //当 Spinner 的某一条数据被选中时回调的方法
            @Override
            public void onItemSelected(AdapterView<?> adapterView, View view, int position, long id) {
                temperatureHDFragment.this.position = position;
                temperatureHDFragment.this.hasRunned = false;
                if (position != 0) {
                    lineChartView.setVisibility(View.GONE);
```

```
                    textDataTip.setText("在这里显示历史数据");
                    textDataTip.setVisibility(View.VISIBLE);
                }
                reflectChangeSpinnerPosition(AdapterView.INVALID_POSITION);
            }
            //当 Spinner 的数据没有被选中时回调的方法
            @Override
            public void onNothingSelected(AdapterView<?> adapterView) {
            }
        });
    }
    /*初始化用户的 ID 和 KEY 以及使用的智云服务器地址*/
    private void initSetting(){
        String id;
        String key;
        String serverAddress;
        id = preferences.getString("id",config.getUserId());
        key = preferences.getString("key",config.getUserKey());
        serverAddress = preferences.getString("server" + ":8080",config.getServerAddress()+ ":8080");
        wsnHistory.setIdKey(id,key);
        wsnHistory.setServerAddr(serverAddress);
    }
    /*通过 Java 反射来让 Spinner 选择同一个选项时触发 onItemSelected 事件*/
    private void reflectChangeSpinnerPosition(int position){
        try {
            Field field = AdapterView.class.getDeclaredField("mOldSelectedPosition");
            field.setAccessible(true);    //设置 mOldSelectedPosition 可访问
            field.setInt(airHumiditySearch,position); //设置 mOldSelectedPosition 的值
        } catch (Exception e) {
            e.printStackTrace();
        }
    }
    /*该方法用于更新 UI*/
    private void mainThreadUpdateUI(final String json){
        getActivity().runOnUiThread(new Runnable() {
            @Override
            public void run() {
                lineChartView.setVisibility(View.VISIBLE);
                initLineChart(json);
                textDataTip.setVisibility(View.GONE);
            }
        });
    }
    private void onSpinnerItemSelected(final int position) {
        hasRunned = true;
        if (position == 1) {
            new Thread(new Runnable() {
```

```
                @Override
                public void run() {
                    try {
                        String result = wsnHistory.queryLast1H(airHumidityMac + "_A1");
                        mainThreadUpdateUI(result);
                    } catch (Exception e) {
                        e.printStackTrace();
                    }
                }
            }).start();
        } else if (position == 2) {
            new Thread(new Runnable() {
                @Override
                public void run() {
                    try {
                        String result = wsnHistory.queryLast6H(airHumidityMac + "_A1");
                        mainThreadUpdateUI(result);
                    } catch (Exception e) {
                        e.printStackTrace();
                    }
                }
            }).start();
        } else if (position == 3) {
            new Thread(new Runnable() {
            @Override
            public void run() {
                try {
                    String result = wsnHistory.queryLast12H(airHumidityMac + "_A1");
                    mainThreadUpdateUI(result);
                } catch (Exception e) {
                    e.printStackTrace();
                }
            }
        }).start();
    } else if (position == 4) {
        new Thread(new Runnable() {
            @Override
            public void run() {
                try {
                    String result = wsnHistory.queryLast1D(airHumidityMac + "_A1");
                    mainThreadUpdateUI(result);
                } catch (Exception e) {
                    e.printStackTrace();
                }
            }
        }).start();
    } else if (position == 5) {
```

```
                new Thread(new Runnable() {
                    @Override
                    public void run() {
                        try {
                            String result = wsnHistory.queryLast5D(airHumidityMac + "_A1");
                            mainThreadUpdateUI(result);
                        } catch (Exception e) {
                            e.printStackTrace();
                        }
                    }
                }).start();
            } else if (position == 6) {
                new Thread(new Runnable() {
                    @Override
                    public void run() {
                        try {
                            String result = wsnHistory.queryLast14D(airHumidityMac + "_A1");
                            mainThreadUpdateUI(result);
                        } catch (Exception e) {
                            e.printStackTrace();
                        }
                    }
                }).start();
            } else if (position == 7) {
                new Thread(new Runnable() {
                    @Override
                    public void run() {
                        try {
                            String result = wsnHistory.queryLast1M(airHumidityMac + "_A1");
                            mainThreadUpdateUI(result);
                        } catch (Exception e) {
                            e.printStackTrace();
                        }
                    }
                }).start();
            } else if (position == 8) {
                new Thread(new Runnable() {
                    @Override
                    public void run() {
                        try {
                            String result = wsnHistory.queryLast3M(airHumidityMac + "_A1");
                            mainThreadUpdateUI(result);
                        } catch (Exception e) {
                            e.printStackTrace();
                        }
                    }
                }).start();
```

```
} else if (position == 9) {
    new Thread(new Runnable() {
        @Override
        public void run() {
            try {
                String result = wsnHistory.queryLast6M(airHumidityMac + "_A1");
                mainThreadUpdateUI(result);
            } catch (Exception e) {
                e.printStackTrace();
            }
        }
    }).start();
} else if (position == 10) {
    new Thread(new Runnable() {
        @Override
        public void run() {
            try {
                String result = wsnHistory.queryLast1Y(airHumidityMac + "_A1");
                mainThreadUpdateUI(result);
                Log.d("sss", result);
            } catch (Exception e) {
                e.printStackTrace();}
            }
        }).start();
    }
}
@Override
public void onStart() {
    super.onStart();
    reflectChangeSpinnerPosition(position);
}
@Override
public void onDestroyView() {
    lcApplication.unregisterOnWSNDataListener(this);
    super.onDestroyView();
}
private void initLineChart(String json) {
    if (json == null) {
        lineChartView.setVisibility(View.GONE);
        textDataTip.setText("服务器目前没有数据");
        textDataTip.setVisibility(View.VISIBLE);
    } else {
        List<PointValue> mPointValues = new ArrayList<>();
        List<AxisValue> mAxisXValues = new ArrayList<>();
        List<HistoricalData.DatapointsBean> datapointsBeanList;
        HistoricalData historicalData = new HistoricalData();
        try {
```

```
JSONObject jsonObject = new JSONObject(json);
String current_value = jsonObject.optString("current_value");
JSONArray datapoints = jsonObject.optJSONArray("datapoints");
datapointsBeanList = new ArrayList<>();
for (int j = 0; j < datapoints.length(); j++) {
    HistoricalData.DatapointsBean datapointsBean = new HistoricalData.DatapointsBean();
    JSONObject jsonObject1 = datapoints.optJSONObject(j);
    String at = jsonObject1.optString("at");
    String value = jsonObject1.optString("value");
    datapointsBean.setAt(at);
    datapointsBean.setValue(value);
    datapointsBeanList.add(datapointsBean);
    float v = Float.parseFloat(value);
    mAxisXValues.add(new AxisValue(j).setLabel(at.substring(0, 10) + " " +at.
                                    substring(11, 19)));
    mPointValues.add(new PointValue(j, v).setLabel(value));
}
historicalData.setCurrent_value(current_value);
historicalData.setDatapoints(datapointsBeanList);
Line line = new Line(mPointValues).setColor(getActivity().getResources().getColor(R.
                color.btn_home_pager_background));  //折线的颜色（橙色）
line.setShape(ValueShape.CIRCLE);//折线图上每个数据点的形状
//这里是圆形（有三种：ValueShape.SQUARE、ValueShape.CIRCLE、ValueShape.
 DIAMOND）
line.setCubic(false);              //曲线是否平滑，是曲线还是折线
line.setFilled(false);             //是否填充曲线
line.setHasLabels(true);           //曲线的数据坐标是否加备注
line.setHasLines(true);            //是否用线显示，如果为 false 则没有曲线，只有点
line.setHasPoints(true);
//是否显示原点，如果为 false 则没有原点，只有点显示（每个数据点都是大圆点）
line.setHasLabelsOnlyForSelected(true);
line.setHasLabelsOnlyForSelected(true);
//单击数据坐标提示数据（设置了 line.setHasLabels(true)则无效）
List<Line> lines = new ArrayList<>();
lines.add(line);
LineChartData data = new LineChartData();
data.setLines(lines);
Axis axisX = new Axis();                    //X 轴
//X 坐标轴字体是斜的显示还是正的，true 表示字体是斜的
axisX.setHasTiltedLabels(false);
axisX.setTextColor(Color.BLUE);             //设置字体颜色
axisX.setName("");                          //表格名称
axisX.setTypeface(Typeface.DEFAULT_BOLD);
axisX.setAutoGenerated(true);
axisX.setTextSize(10);                      //设置字体大小
axisX.setMaxLabelChars(5);
axisX.setValues(mAxisXValues);              //填充 X 轴的坐标名称
```

```
                data.setAxisXBottom(axisX);                    //X 轴在底部
                axisX.setHasLines(true);                       //X 轴分割线
                Axis axisY = new Axis();                       //Y 轴
                axisY.setName("");                             //Y 轴标注
                axisY.setTextColor(Color.BLUE);                //设置字体颜色
                axisY.setAutoGenerated(true);
                axisY.setTypeface(Typeface.DEFAULT_BOLD);
                axisY.setInside(true);
                data.setAxisYLeft(axisY);                      //Y 轴设置在左边
                //设置行为属性，支持缩放、滑动以及平移
                lineChartView.setInteractive(true);
                lineChartView.setZoomType(ZoomType.HORIZONTAL);
                lineChartView.setMaxZoom((float) 800);
                lineChartView.setContainerScrollEnabled(true, ContainerScrollType.HORIZONTAL);
                lineChartView.setLineChartData(data);
                lineChartView.setVisibility(View.VISIBLE);
                lineChartView.setZoomLevel(100,0,500);
            } catch (JSONException e) {
                e.printStackTrace();
                lineChartView.setVisibility(View.GONE);
                textDataTip.setText("数据解析出现异常");
                textDataTip.setVisibility(View.VISIBLE);
            }
        }
    }
    //当智云服务器消息到达时回调的方法
    @Override
    public void onMessageArrive(String mac, String tag, String val) {
        if (airHumidityMac == null) {
            wsnrtConnect.sendMessage(mac, "{TYPE=?}".getBytes());
        }
        if (tag.equalsIgnoreCase("TYPE") && val.substring(2, val.length()).equals("601")) {
            airHumidityMac = mac;
        }
        if (airHumidityMac != null&&position!=0) {
            if (!hasRunned) {
                onSpinnerItemSelected(airHumiditySearch.getSelectedItemPosition());
            }
        }
    }
    //当程序与服务器断开连接时回调的方法
    @Override
    public void onConnectLost() {
    }
    //当程序与服务器建立连接时回调的方法
    @Override
    public void onConnect() {
```

```
        }
    }
```

2．仓库环境管理系统硬件部署

（1）设备选型。仓库环境管理系统硬件主要使用经典型无线节点 ZXBeeLiteB、采集类开发平台（Sensor-A）、控制类开发平台（Sensor-B）、Android 网关。仓库环境管理系统的硬件部署如图 19.11 所示。

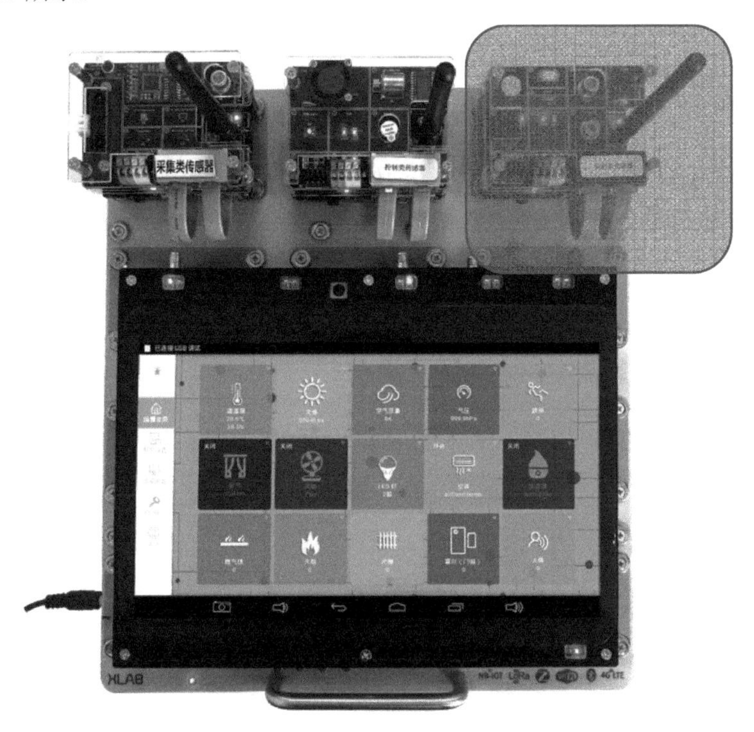

图 19.11　仓库环境管理系统的硬件部署

（2）设备配置。

① 连接硬件，固化 Android 网关、ZigBee 协调器和经典型无线节点的程序。

② 修改 Android 网关、ZigBee 协调器和经典型无线节点的网络参数，正确设置 Android 网关，将 ZigBee 网络接入智云物联平台。

（3）设备组网。

① 构建 ZigBee 网络，并让传感器节点正确接入 ZigBee 网络。

② 启动 Android 网关，观察节点是否正确入网。

③ 通过综合测试软件查看网络拓扑结构。

（4）安装应用软件。连接 Android 网关与 PC，将仓库环境管理系统的应用程序 WarehouseManagement.apk 安装到 Android 网关。仓库环境管理系统的应用程序安装成功后的界面如图 19.12 所示。

图 19.12 仓库环境管理系统的应用程序安装成功后的界面

19.5 任务验证

在 Android Studio 开发环境中打开任务的例程，编译通过后运行程序。

首先在“更多信息”页面设置智云服务器的 ID、KEY，如图 19.13 所示。

图 19.13 设置智云服务器的 ID、KEY

如果连接成功智云服务器，则 MAC 设置中会显示传感器节点的 MAC 地址，如图 19.14 所示。

图 19.14 在 MAC 设置中显示节点的 MAC 地址

切换到“运营首页”页面，这时会显示仓库环境管理系统的主界面，如图 19.15 所示。

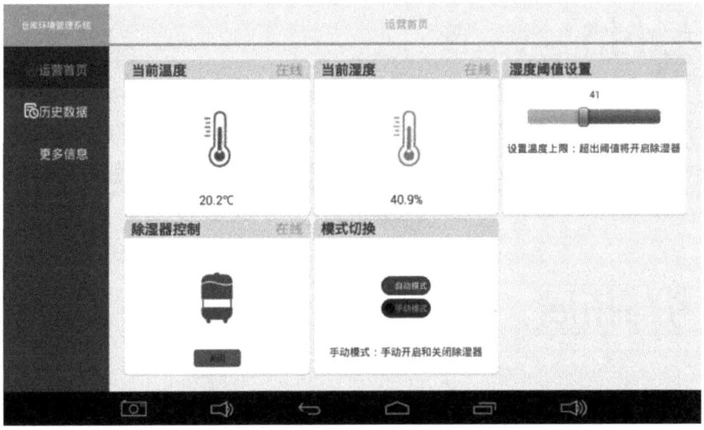

图 19.15　仓库环境管理系统的主界面

在“历史数据”页面中，可通过选择时间段来查看不同时间段的数据折线图，如图 19.16 和图 19.17 所示。

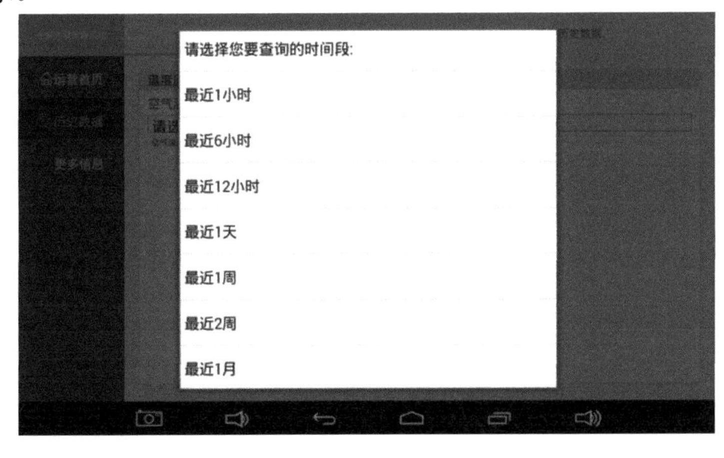

图 19.16　选择要查询的时间段

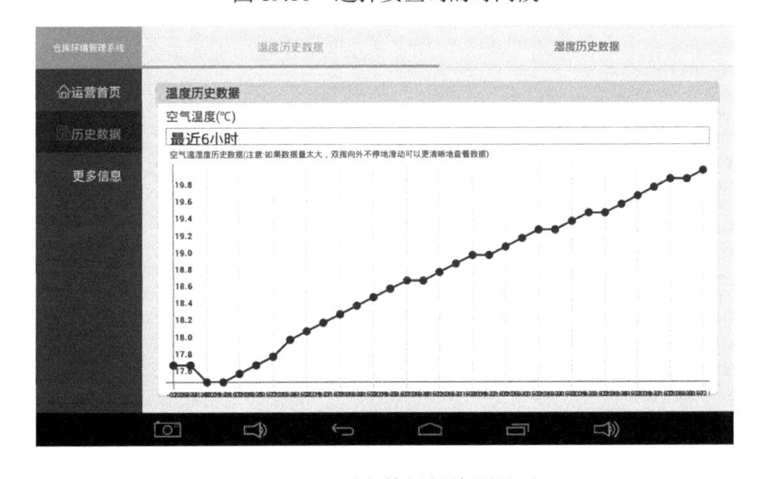

图 19.17　历史数据折线图显示

19.6 开发小结

本任务以仓库环境管理系统为例，介绍了项目开发的总体架构、系统功能、系统通信流程，通过智云物联平台和 Android 应用程序接口实现了环境温湿度的实时监测，以及历史数据的查询功能。

19.7 思考与拓展

（1）本任务使用了哪些应用程序接口？

（2）智云物联平台提供的 Android 开发接口有哪些？分别有什么功能？

参考文献

[1] 李刚. 疯狂 Java 讲义[M]. 5 版. 北京：电子工业出版社，2019.

[2] 明日科技. Java 从入门到精通[M]. 5 版. 北京：清华大学出版社，2019.

[3] 李兴华. Java 从入门到项目实战[M]. 北京：中国水利水电出版社，2019.

[4] 凯・S.霍斯特曼. Java 核心技术 卷 I：基础知识[M]. 林琪，苏钰涵，等译. 北京：机械工业出版社，2016.

[5] 欧阳燊. Android Studio 开发实战：从零基础到 App 上线[M]. 2 版. 北京：清华大学出版社，2018.

[6] 刘望舒. Android 进阶之光[M]. 北京：电子工业出版社，2017.

[7] 李刚. 疯狂 Android 讲义[M]. 4 版. 北京：电子工业出版社，2019.

[8] 安辉. Android App 开发从入门到精通[M]. 北京：清华大学出版社，2018.

[9] 明日学院. Android 开发从入门到精通（项目案例版）[M]. 北京：中国水利水电出版社，2017.

[10] 廖建尚. 面向物联网的传感器应用开发技术[M]. 北京：电子工业出版社，2019.

[11] 赵飞，叶震. UDP 协议与 TCP 协议的对比分析与可靠性改进[J]. 计算机技术与发展，2006，(09):219-221.

[12] 廖建尚. 物联网开发与应用——基于 ZigBee、Simplici TI、低功率蓝牙、Wi-Fi 技术[M]. 北京：电子工业出版社，2017.

[13] 在 Android 平台上构建任何应用[EB/OL]. [2019-12-11]. https://developer.android.google.cn.

[14] JavaScript 教程[EB/OL]. [2020-2-11]. https://www.runoob.com/js/js-tutorial.html.

[15] Android 基础入门教程[EB/OL]. [2019-12-17]. https://www.runoob.com/w3cnote/android-tutorial-intro.html.

[16] 安卓学习笔记[EB/OL]. [2019-12-20]. http://android.52fhy.com/java/index.html.

[17] 一看就懂的 Android APP 开发入门教程[EB/OL]. （2014-5-26）[2019-12-20]. https://www.jb51.net/article/50395.htm.

[18] 为何要用 Fragment[EB/OL]. （2018-7-27）[2019-12-30]. https://www.jianshu.com/p/c55abd4c7140.

[19] MR_Codingson. SharedPreferences 的用法以及详解[EB/OL].（2017-2-28）[2019-12-30]. https://blog.csdn.net/MR_Condingson/article/details/58586419.

[20] wminxue319. Android 中的图形图像[EB/OL]. （2012-2-16）[2019-12-30]. https://blog.csdn.net/wminxue319/article/details/7263603.

[21] 极客学院. Android 工程师[EB/OL]. [2020-2-10] . http://ke.jikexueyuan.com/zhiye/android.

[22] 兮若. Java 基础-第三章（流程语句）[EB/OL].（2018-2-26）[2020-2-15]. https://my. oschina.

net/jason26/blog/1623907.

[23] heyuchang666. Android 基本 UI（二）——Button、ImageButton[EB/OL].（2014-12-8）[2020-2-17]. https://blog.csdn.net/heyuchang666/article/details/41802893.

[24] djw_920506. 控件 Spinner 的用法[EB/OL].（2015-11-2）[2020-2-20]. https://blog.csdn.net/u013925475/article/details/49583907.

[25] Android 事件传递机制（按键事件）[EB/OL].（2014-07-30）[2020-2-23]. https://blog.csdn.net/colorapp/article/details/38299315.

[26] heyuchang666. Android 事件和监听器详细的介绍[EB/OL].（2014-12-5）[2020-2-27]. https://blog.csdn.net/heyuchang666/article/details/41748645.

[27] Android 事件模型[EB/OL].（2019-02-07）[2020-3-5]. http://www.doc88.com/p-7843820673664.html.

[28] Activity 生命周期详解[EB/OL].（2017-09-07）[2020-3-9]. https://blog.csdn.net/ u011941673/article/details/77880457.

[29] android 第一行代码－3.activity 之间的调用跟数据传递[EB/OL]. [2020-3-12]. https://www.cnblogs.com/alexkn/p/5449516.html.

[30] 航天飞哥. Android之服务（五）IntentService的使用[EB/OL].（2016-01-27）[2020-3-9]. https://blog.csdn.net/liu857279611/article/details/50594869.

[31] Wawdjr19970078. 安卓四大组件之一广播[EB/OL].（2018-03-26）[2020-3-17]. https://blog.csdn.net/wawdjr19970078/article/details/79699649.

[32] Carson_Ho. Android：关于 ContentProvider 的知识都在这里了！[EB/OL].（2017-07-26）[2020-3-20]. https://blog.csdn.net/carson_ho/article/details/76101093.

[33] fragment-Android 6.0 开发者文档[EB/OL].（2018-04-13）[2020-3-27]. https://blog.csdn.net/qq_20802379/article/details/79924111.

[34] Android 资源文件 res 的使用详解（strings、layout、drawable、arrays、动画等）[EB/OL].（2015-07-14）[2020-3-30]. https://blog.csdn.net/jian_csdn/article/details/46875105.

[35] Android 数据库的使用——Sqlite[EB/OL].（2018-2-14）[2020-4-3]. https://www.jianshu.com/p/1ef7cfb69cde.

[36] SQLiteOpenHelper 使用详解[EB/OL].（2016-09-05）[2020-4-8]. https://blog.csdn.net/zzq123686/article/details/52439800.

[37] Android 数据存储的五种方法汇总[EB/OL]. [2020-4-12]. https://www.cnblogs.com/chengzhengfu/p/4582515.html.

[38] Sunzxyong. Android 中 Socket 通信之 TCP 与 UDP 传输原理[EB/OL].（2015-03-26）[2020-4-16]. https://blog.csdn.net/u010687392/article/details/44649589.